TRANSACTIONS

OF THE

AMERICAN PHILOSOPHICAL SOCIETY

HELD AT PHILADELPHIA
FOR PROMOTING USEFUL KNOWLEDGE

NEW SERIES—VOLUME 66, PART 3
1976

CRYSTALS AND COMPOUNDS

Molecular Structure and Composition in Nineteenth-century French Science

SEYMOUR H. MAUSKOPF
Associate Professor of History, Duke University

THE AMERICAN PHILOSOPHICAL SOCIETY
INDEPENDENCE SQUARE
PHILADELPHIA

July, 1976

Copyright © 1976 by The American Philosophical Society

Library of Congress Catalog
Card Number 76-3197
International Standard Book Number 0-87169-663-0
US ISSN 0065-9746

FOREWORD

My interest in the development of theories of material structure was aroused in graduate seminars at Princeton led by Professor C. C. Gillispie. I owe him a debt beyond words for the direction, encouragement, and friendship he has provided throughout my graduate days and in the years since.

Hardly less do I owe Professor Henry Guerlac, who was the original inspiration to me, as to so many others, to pursue the study of the history of science. I wish especially to thank Professor Guerlac and Professor L. Pearce Williams for their hospitality which enabled me to spend a most pleasant and profitable summer at Cornell University doing research in the splendid collection in the history of chemistry.

I wish also to give especial thanks to Dr. Edwin Clarke, director of the Wellcome Institute for the History of Medicine and formerly head of the Sub-Department of the History of Medicine, University College London, for the warm hospitality extended to me on several occasions. Of the numerous libraries I have used, I am most indebted to the Trustees of the British Museum, and to the authorities of the Bibliothèque Nationale, the latter especially for making available to me in copious quantity, the notebooks and early manuscripts of Louis Pasteur. In this connection, I am also greatly indebted to Professor Wyart of the University of Paris for his generosity in securing for me microfilms of the laboratory notebook in which Pasteur had set down his great discovery of the enantiomorphism of the tartrates.

Duke University has been my home for a decade now, and I am deeply grateful for the liberal opportunities provided here to pursue and complete my research as well as for the Research Council grants which so greatly facilitated my labors. Special praise is due Mrs. Dorothy Sapp of the Duke history department secretarial staff for her speedy yet careful typing of the manuscript.

Of those who have given generously of their time and scholarly acumen in reading the manuscript in its various stages of completion, I wish to single out Professor Satish Kapoor of the University of Saskatchewan at Regina, Dr. Betty Jo Dobbs of Northwestern University, Professor Gert Brieger formerly of Duke University, and now Chairman of the Department of the History of Health Sciences, University of California, San Francisco, and my especially close friend and colleague, Professor Michael McVaugh of the University of North Carolina.

Lastly, to express the unexpressable: a word for my wife, whose enthusiasm, encouragement and honest criticism have sustained me at all stages of this study.

S. H. M.

CRYSTALS AND COMPOUNDS
Molecular Structure and Composition in Nineteenth-century French Science

Seymour H. Mauskopf

CONTENTS

	PAGE
Introduction	5
I. Molecular form and crystal structure	7
II. The molecular "individual"	14
III. Mineral species and quantified chemistry	21
IV. Molecular geometry	33
V. Physical optics and the molecule	55
VI. Pasteur's discovery of enantiomorphism: the synthesis of a tradition	68
Conclusion	79
Index	81

INTRODUCTION

Recent scholarship has greatly extended our view of the history of theories about the ultimate nature and structure of matter. This is particularly true for the period before the genesis of Dalton's atomic theory. We now know something about the hierarchical theories of material structure prevalent in the seventeenth century, in the thought of Boyle and Newton, for example. We have begun to perceive how important Newton's ideas on the structure of matter and on the forces which bound substances together or pulled them apart were for physical and chemical philosophers of the eighteenth century. Dalton's chemical atomic theory no longer seems quite so bereft of background as it did twenty-five years ago. At the same time, the real innovation of Dalton in turning from consideration of force to consideration of atomic weight has also been clarified.

For the post-Daltonian period, the variegated historical picture of nineteenth-century theorizing on the nature of matter has not yet emerged with clarity of pre-Daltonian Newtonianism. But at least we do have now a good idea of the development of a few of the main themes, some of which were not apparent before recent scholarship focused upon them. We know of the tradition of resistance to Daltonian atoms which spanned the whole of the nineteenth century, of the perseverance of interest in the dynamics of chemical combination, and we have some idea of the theoretical issues connected with the rise of organic chemistry, to take some examples.

With the filling in of the eighteenth-century background to Dalton's chemical atomic theory and of the nineteenth-century context of its elaboration and influence, there was one approach to the understanding of the intimate structure of matter which also developed in this period but which has been neglected by historians of science. I refer to the attempt to get at the forms of molecules through the study of crystal structure. The major figure connected with this approach was the Abbé René Just Haüy (1743–1822), whose molecular theory of crystal structure was the first comprehensive mathematical treatment of that subject.

Despite the enormous importance of crystallography to modern physics and chemistry, the history of this subject has received relatively scanty attention and its integration into the general history of physical science has not progressed very far. Haüy has fared better than most crystallographers in the attention bestowed upon him, but even he and his work have remained isolated and outside the general considerations of pre-Daltonian material speculations. The development of crystallography in the nineteenth century has, to date, received only sketchy treatment and there have been only a few studies which have even touched upon the relationship between crystallography and other physical sciences in this period.

My own interest in this subject was sparked by my study of one of the most fascinating and important discoveries in the nineteenth century: Louis Pasteur's discovery during the spring of 1848 of enantiomorphism—left-handed and right-handed asymmetry—in tartrate crystals. This was Pasteur's first major scientific achievement. It instantly secured the young chemist a high place among his contemporary scientists. It conditioned the course of much of his subsequent scientific work. It eventually led the way to the development of stereochemistry.

Yet it has been all but ignored by historians of science. Even in the one serious study it has received, by J. D. Bernal, the historiographical isolation of crystallography from other physical sciences was emphasized. In this discovery, chemical, optical, and crystallographical theories and data were brought into an extraordinary synthesis by Pasteur. Recognizing the importance of this for nineteenth-century science, Bernal wrote:

In the account of this story as it appears in the biographies of Pasteur or in general books on the history of science, the main accent is on the brilliance and intuition of the young genius. Enough account is not taken of the fact that Pasteur's discovery was entirely logical and held, from the very start, a central place in the history of science.

However, he went on to write:

It arose like all the great discoveries of science, at the meeting place of hitherto distinct disciplines.[1]

In the course of my own research it became clear to me that Pasteur's discovery resulted not so much from the intersection of *distinct* disciplines as from a well-established French tradition of *interaction* between them. It is this interaction which I treat in this study, focusing on the period from Haüy's first elaboration of his molecular crystal structure theory to Pasteur's discovery of enantiomorphism. This period is a particularly important one for each of the interacting sciences; each can be said to have assumed its recognizably modern guise in these years. For crystallography, the period encompasses the work of Haüy at the beginning and the mathematical theory of crystal lattice structure developed by Auguste Bravais (as well as Pasteur's discovery) at the end. For chemistry, this period witnessed the implementation of Lavoisier's gravimetric and pneumatic chemistry, the appearance of Dalton's atomic theory, the development of electrochemistry, and the rise of organic chemistry. It was in these years, finally, that physical optics, relatively dormant in the eighteenth century, became an important branch of the physical sciences, particularly with the discovery and study of polarization phenomena.

The guiding theme in my study of this scientific interaction is the issue of "transdiction."[2] By this term, I mean the inferential procedure whereby the properties of matter at the level of the imperceptible are elucidated by studying and analyzing material properties at the macroscopic level. The particular assumption common to the scientists whose work and theories are to be examined here is what I shall call a molecular assumption: all material bodies are composed of discrete submicroscopic material units—molecules—which have the same essential properties as the macroscopic body of which they are the components. Molecules are formed by chemical synthesis; they then aggregate together, without losing their discreteness, to make up the macroscopic body. Moreover, the molecules of each substance are invariant in their properties. The transdictive problem was how to specify these properties.

It was in Haüy's theory of crystal structure that the molecular assumption, as I have defined it, became of central importance to a theory for the first time. For in his theory, Haüy outlined a transdictive procedure for specifying the invariant forms of the molecule. By implication, as realized by Haüy in his mineralogical considerations, formal invariance also meant invariance in composition. This raised an issue which contemporary chemists were just then debating.

In focusing upon the molecule as the invariant unit of composition, Haüy and his supporters of around 1800 were anticipating the turn which chemistry was to take with Dalton's atomic theory. The molecular concept had always been a chemical concept, but before Dalton it had not been central to chemical theory. Dalton's theory, examined from a transdictive point of view, had many parallels with Haüy's molecular theory. Dalton's achievement was to unpack the compositional aspect of the molecule beyond the mere assertion of invariance, to show what this invariance meant in terms of numbers of component atoms. His instrumental concept was relative atomic weight; the power of his theory lay, like Haüy's, in demonstrating mathematical patterns in otherwise variegated data. In both theories, assumptions of simplicity were necessary in order to effect the transdictive procedure: in Haüy's case these had to do with the manner in which he envisioned molecular structuring; in Dalton's with assumptions about the numbers of atoms which made up the molecule.

Haüy had conceived of his molecules as small polyhedra, composed, in turn, of determinately shaped elementary polyhedra (which latter shapes, however, he never really tried to ascertain). Dalton conceived of his atoms as spheres. But from the time of the inception of Dalton's theory it was felt by some to be necessary to specify not only the numbers of atoms in a molecule but their arrangement as well and in France, especially, there developed something of a programmatic tradition of combining Daltonian chemical atomism with Haüyian theory of crystal structure for their mutual benefit. The location of this in France was natural enough, for Haüy's crystallography maintained a dominant position there during Haüy's lifetime and afterwards which it did not enjoy elsewhere.

The rising interest, mainly in France, in physical optics also added another component to the transdictive problem of the nature of the molecule. What did the behavior of light in macroscopic materials tell about the optical properties of their molecules? What could one infer about the nature of molecules—their composition and form—from their optical properties? Polarization phenomena, as studied by one scientist in particular, the physicist Jean Baptiste Biot, were the focus of these questions.

---

[1] J. D. Bernal, "Molecular Asymmetry," *Science and Industry in the Nineteenth Century* (Bloomington, Indiana University Press, 1970), p. 182.

[2] *Cf.* J. E. McGuire, "Atoms and the 'Analogy of Nature': Newton's Third Rule of Philosophizing," *Studies in History and Philosophy of Science* 1 (1970): pp. 3–58 for an analysis of seventeenth-century atomism in terms of this concept (called "transduction" by McGuire) and H. Guerlac, "The Background to Dalton's Atomic Theory," *John Dalton and the Progress of Science*, ed. D. S. L. Cardwell (Manchester, Manchester University Press, 1968), pp. 64 ff. for Newton and the eighteenth century.

In the course of his education at the École Normale, Louis Pasteur came into fruitful contact with three scientists, each of whom had been grappling for years with the problems of molecular structure and composition. They were the chemist Auguste Laurent, the crystallographer Gabriel Delafosse, and Biot. The impact of their ideas and approaches upon the young scientist led to his great discovery of enantiomorphism. By good fortune, we are able to follow the complex development of Pasteur's own thoughts and experimental interests up to the very moment of his discovery through the rich material which has survived from his student days and from the time of his first major research.

Crystallography, chemistry, and optics together have finally provided the key to unlock the secrets of atomic arrangement in this century. Pasteur and his predecessors were not able to get this far, but the goal of explicating molecular structure and composition was nevertheless an important and fruitful one in nineteenth-century science. Haüy's molecular theory of crystal structure will provide the starting point for our investigation into the complex scientific interaction of the early nineteenth century in pursuit of this goal.

I. MOLECULAR FORM AND CRYSTAL STRUCTURE

In 1811 John Dalton published a paper in *Nicholson's Journal* in which he surveyed some of the definitions and uses of terms like "particle" and "molecule" current in the chemical literature. Dalton's point was to criticize nearly all the writers he examined in the light of his own recently developed chemical atomic theory for their vague or unhelpful definitions of these theoretical terms. Only one definition received his full approbation: that of "integrant particle" by John Murray. This Dalton deemed to be "the best definition I have yet seen of *integrant particle*." Murray's definition was as follows:

The calcareous spar, to take it as an example, may be reduced to a particle beyond which, if its minuteness allowed us to operate upon it, it is demonstrable its figure would not change. To these last particles, the result of the mechanical analysis, Haüy gives the name of *integrant particles*, and their union constitutes the crystal.[1]

It is clear from Dalton's selection that what he was interested in was a good definition of the chemical molecule—that smallest unit of a substance whose chemical composition was still intact, but whose further division would decompose it into its constituent elements. What is striking about Murray's definition is that it included not only this compositional aspect, but also the aspect of structure or form. And by Murray's own testimony, the formal aspect had already been delineated in the crystal structure theory of René Just Haüy, the founder of modern crystallography.

Haüy, an abbé and teacher at the Collège Cardinal Lémoine, had come late to the study of crystals, by way of an interest in botanical classification. He was over forty when he published his first detailed account of the crystal structure theory, the *Essai d'une théorie sur la structure des cristaux,* in 1784. Arguably, it was only in the late eighteenth century that a distinct science of crystallography became discernable. Up until then, the study of crystals was a part of natural history, of mineral taxonomy, of geology, or of chemistry. Crystallography was to maintain intimate connections with these other sciences, but with Haüy, building upon the work of his two notable predecessors, the naturalist Jean Baptiste de Romé de l'Isle and the chemist, Torbern Bergman, it developed an identity of its own.

The fundamental problem in the study of crystals was that they could and did crystallize in a great variety of shapes. The first step in developing a science of crystallography was to establish the nature of the regularities which underlay the diversity of crystalline shape. It was Romé de l'Isle and Bergman who first began to determine systematically what these regularities were. Haüy took the next step: the construction of a comprehensive and consistent theory of crystal structure which utilized these regularities. His theory was based on a polyhedral molecular unit of structure, the *molécule intégrante,* or Murray's "integrant particle."

Crystals and gemstones had always exerted great fascination by virtue of their regular form and their luster. By the mid-eighteenth century, if there was as yet no comprehensive theory of crystal structure, there were current a number of different approaches to explaining crystal growth and form, with no one of them, however, dominant and with naturalists able to combine several without much sense of incompatibility. There were still some who thought that minerals in general and crystals in particular "grew" from a form-giving germ or that crystals in some way received their forms from organisms. More chemically oriented was the belief that crystals owed their shapes to some kind of formal agent or form-giving "principle" which was a necessary constituent of every crystalline substance. "Salt" in its Paracelsian sense as the principle of solidity and structure was the formal agent most frequently cited in this connection.

Side by side with these were more strictly materialistic explanations. In the seventeenth century, Robert Hooke and Christiaan Huyghens had derived models for Iceland spar based upon the packing together of spherical or spheroidal corpuscles. Others in the eighteenth century, notably Torbern Bergman, made use of the supposition that crystals grew by the accre-

---

[1] J. Dalton, "Inquiries Concerning the Signification of the Word Particle, as Used by Modern Chemical Writers, as well as Concerning Some Other Terms and Phrases," *Nicholson's Journal* (A Journal of Natural Philosophy, Chemistry and the Arts) **28** (1811): p. 85.

tion of very thin sheets of material on each crystalline face.[2] Haüy himself used this idea in his first papers on crystal structure which he had published shortly before the *Essai d'une théorie*.[3]

Finally, there was the supposition that crystals were composed of polyhedral molecules. This notion was also a venerable one and it had received some support in microscopal studies of crystal formation and growth. This was particularly striking in the case of common salt which the chemist G. F. Rouelle studied in the 1740's. The cubic crystalline form of salt was discernable in the very smallest salt crystals and even in ill-formed ones.[4]

The idea of a polyhedral molecular unit of crystal structure had been adopted by the French chemist, P. J. Macquer and expounded in his popular handbook, the *Dictionnaire de chymie* of 1766. Early in this dictionary, in the article "aggregation," Macquer had given one of the first really clear definitions of the "molecule" as a unit of composition:

By the integrant parts of a body are to be understood the smallest molecules or particles into which this body can be reduced without decomposition. We may conceive that a neutral salt, for instance common salt, may be divided into molecules still smaller and smaller, without separation of the acid and alkali which constitute the salt; so that these molecules, however small, shall always be common salt, and possessed of all its essential properties. If we should now suppose that these molecules are arrived at their utmost degree of smallness, so that each of them shall be composed of one atom of acid and of another atom of alkali, and that they cannot be further divided without a separation of the acid and the alkali, then these last molecules are those which Mr. Macquer in his Chemical Lectures calls primitive *integrant molecules*.[5]

Macquer's main concern in this article was to work out a two-level hierarchy into which he believed that material structure generally could be analyzed. On the first level, homogeneous *molécules primitives intégrantes* were formed out of their heterogeneous constituent "atomes" in chemical reaction. These compound molecules then aggregated together to form the second level, the material mass. Macquer had underscored this distinction between the two levels by differentiating between the forces which caused the entities on each level to come into being. The chemical synthesis which produced the compound molecule was brought about by the force of *affinité de composition*. Their aggregation into material mass was caused by an *affinité d'aggrégation*. These forces, although perhaps not intrinsically different, often appeared to act counter to one another.[6]

Macquer also wrote a long and comprehensive article on crystallization in the *Dictionnaire* and in it, he treated crystal formation as a particular type of aggregation. Crystallization was:

... an operation by which the integrant parts of a body, separated from each other by the interposition of a fluid, are disposed to unite again and to form solid, regular, and uniform masses.[7]

Crystals of the same substance always had a regular and invariant figure, the reason being that their integrant molecules also were determinately shaped and symmetrically arranged. This point was brought out with force in the last two of four principles on the "mechanism of crystallization":

3. That although we do not know the figure of the primitive integrant molecules of any body, we cannot doubt but that the primitive integrant molecules of every different body have a constantly uniform and peculiar figure.

4. That the integrant parts cannot have an equal tendency to unite indiscriminately by any of their sides, but by some preferably to others, excepting all the sides of an integrant part of a body be equal and similar; and probably the sides, by which they tend to unite, are those by which they can touch most extensively and immediately.

Therefore:

If, besides, they have time and liberty to unite with each other by the sides most disposed to this union, they will form masses of a figure constantly uniform and similar.[8]

Macquer's principles were good as far as they went, but they failed to come to grips with a fundamental problem of the study of crystal forms: crystals of the same substance do not always have similar figures; they may crystallize in a wide variety of shapes. Was there an ordering principle by which the diverse shapes

---

[2] H. Metzger, *La génèse de la science des cristaux* (Paris, Alcan, 1918) remains the best account of the various theories of crystal growth held in the eighteenth century. *Cf.* also, J. G. Burke, *Origins of the Science of Crystals* (Berkeley, University of California Press, 1966), pp. 10–51.

[3] R. J. Haüy, "Extrait d'une mémoire sur la structure des cristaux de garnet"; "Extrait d'une mémoire sur la structure des spaths calcaires," *Observations sur la physique, sur l'histoire naturelle et sur les arts* 19 (1782): pp. 366–370 and 20 (1782): pp. 33–47 respectively. Hereafter this journal and its successor cited as *Jour. de phys.*

[4] G. F. Rouelle, "Sur le sel marin (première partie). De la crystallisation du sel marin," *Paris, Mém. Acad. Sci.* 1745: pp. 57–79.

[5] P. J. Macquer, *Dictionnaire de chymie* (1st ed., 2 v., Paris, Lacombe, 1766) 1: pp. 55–56. Translation from *A Dictionary of Chemistry* (2 v., London, T. Cadell and P. Elmsly [etc.], 1771) 1: p. 28.

[6] This distinction of affinity forces made in his article, "Affinité," *Dictionary* 1: pp. 48 ff. In the second edition of the *Dictionnaire*, Macquer showed concern for placating the naturalist Buffon, whose unitary principle of chemical affinity his own seemed to contradict (2nd ed., 2 v., Paris, Imprimerie de Monsieur, 1777) 1: p. 40, fn. 1. Henceforth references to the *Dictionnaire* will be to the first edition.

[7] P. J. Macquer, "Cristallisation des sels & d'autres substances," *Dictionnaire* 1: pp. 291–292.

[8] *Ibid.*, pp. 291–292. Translation from *A Dictionary of Chemistry* 1: p. 183. On the basis of Rouelle's work, Macquer thought it was likely that the *molécules primitives intégrantes* of common salt were cubic. *Dictionnaire*, p. 301, fn. (1).

of such crystals could be brought into relation with one another? Macquer himself did not develop his ideas further in this direction, but this question was taken up in most ambitious terms just a few years later.

In 1773 there appeared an *Essai de cristallographie* by Jean Baptiste de Romé de l'Isle (1736–1790). The author of this work had received little formal scientific education; his interest in natural history had developed during military service and travel in India and the Orient. Upon his return to France in 1764 he had been befriended by the mineralogist and metallurgist Balthazar Sage, under whose tutelage his interest had become focused upon mineralogy. Romé de l'Isle subsisted under the patronage of various wealthy friends whose collections of minerals, coins, and gems he supervised and cataloged. Although his scientific work became well known outside of France, he never succeeded in breaking into the French scientific establishment. A proposal for his election to membership in the Académie des Sciences failed and he lived to see his own achievements in mineralogy and crystallography overshadowed by his younger contemporary, Haüy.[9]

In the *Essai de cristallographie,* Romé de l'Isle attempted to make a comprehensive classification of crystals. By the time he wrote the *Essai,* he had made himself extraordinarily familiar with his subject and his work greatly surpassed all its predecessors in scope and detail. To implement his crystal classification, Romé de l'Isle adopted a morphological approach in which he attempted to come to grips with the question that Macquer had begged: how to relate the diverse forms of crystals of the same substance. As a general morphological principle of classification, he introduced the idea of the "primitive form." All crystals of the same substance, no matter how different they appeared to be, had a fundamental and common geometrical form—the primitive form—to which their actual crystalline shapes were somehow related.[10] The source for this idea, and for its justification, lay in two of the already mentioned sets of ideas about the "cause" of crystal form current in the eighteenth century. One was the supposition that crystal form was due to the presence of a particular "saline principle" in the composition of the crystal. The other was the supposition that crystals were built up of polyhedral molecules of specific shapes for each substance. The first of these was for Romé de l'Isle a formal cause in the Aristotelian sense; the second, an efficient cause. This is a good example of the eclecticism in crystallographical explanation common before the appearance of Haüy's theory of crystal structure.

The belief that all crystals contained a "saline principle," while a descendant of the Paracelsian concept of form-giving "salt," had been elaborated and refined since the days of Paracelsus. The vague unitary saline principle had been differentiated into many tangible salts in the seventeenth century. With this had come the attempt to correlate specific crystalline shape with saline composition. Such ideas had been taken up by the Swedish naturalist, Carl Linnaeus, for his classification of the mineral kingdom. Linnaeus had tried to explain the genesis of minerals by means of an analogy with the procreation of organisms. He had suggested that crystals were generated by the "impregnation" of earths by different salts to produce four types of crystalline stones, each with a distinctive crystalline form. All crystalline rocks could be related morphologically (and therefore chemically) to one of these four types.[11] Linnaeus himself did not try to show in any detail how these relationships were to be conceived.

Romé de l'Isle was explicitly attempting to follow in Linnaeus's footsteps in the *Essai de cristallographie*.[12] Reasoning back from form to saline composition, Romé de l'Isle assumed that there must be a common saline ingredient in, for example, such diverse minerals as alum, diamond, iron, and sulfur because their crystal forms were similar. The fact that no salts had been discovered in analyses of stony or metallic crystals did not deter him, in the *Essai* at least, from believing that they would be discovered.[13]

---

[9] "Romé de l'Isle," *Biographie universelle* (52 v., Paris, Michaud frères, etc., 1811–1828) **38**: pp. 521–524. R. Hooykaas, "Romé de l'Isle (or Delisle), Jean Baptiste," Dictionary of Scientific Biography (14 v., N. Y., Scribner's Sons, 1970–) **11**: pp. 520–524.

[10] J. B. L. de Romé de l'Isle, *Essai de cristallographie* (Paris, Didot jeune, 1771), pp. 14 ff. What Romé de l'Isle meant by "primitive form" is well illustrated by his treatment of potassium sulfate (*tartre vitriolé*) crystallizations. The "most perfect form" seemed to be an hexagonal prism terminated at each end by an hexagonal pyramid. Romé de l'Isle then listed ten crystalline varieties of this form, such as one where the two hexagonal pyramids were joined base-to-base without the intermediate hexagonal prism, or one like the foregoing but also with the apices of the pyramids truncated, or one where one pair of opposite sides of the prism and pyramids was almost suppressed, *ibid.*, pp. 52–54. Romé de l'Isle commented on these varieties that, "Toutes ces figures du *Tartre vitriolé*, font voir de combien de formes différentes un même sel est susceptible. Cependant ces dix variétés tendent à former le prisme hexagone & pyramidal de la première figure." *Ibid.*, p. 54.

[11] The four types were niter, muria, natrum and alum. C. Linnaeus, *Systema Naturae* (12th ed., 3 v., Holmiae, L. Salvium, 1766–1768) **3**: pp. 3–7. A fuller exposition of Linnaeus's ideas on this subject is found in M. Kaehler, "Crystallorum Generatio," *Amoenitates Academicae* (10 v., Holmiae, Godofredum Kieswetter, 1749–1790) 1: (1749), XV: pp. 454–482.

[12] J. B. L. de Romé de l'Isle, *op. cit.,* pp. xii, 8, 34–35, 40–42.

[13] Since quartz seemed to have virtually the same primitive form as that of *tarte vitriolé*, a salt, Romé de l'Isle hazarded the suggestion that this latter was the formal agent in quartz crystallization. *Ibid.*, p. 176. "Salt" itself was an expression of a more fundamental form-giving constituent of crystals: water. The various salts which endowed crystals with their forms were compound substances—"élémens secondaires" as he called them—made up of acids and bases which, in turn, were composed of water and some kind of vitrifiable earth in various pro-

The saline principle was linked up to the "primitive form" by Romé de l'Isle's supposition that each salt had its own peculiar primitive form. Moreover, this form was dependent on the form of that salt's *molécule intégrante*. And the efficient cause for crystallization was the regular aggregation of these polyhedral molecules.[14] There is evidence for the influence of Macquer's ideas in the *Essai*, for Romé de l'Isle's first reference to molecular aggregation into crystals was by way of a quotation from the *Dictionnaire* article on crystallization and he subsequently elaborated on this by paraphrasing Macquer's already-quoted principles of crystal formation.[15] The context of this paraphrase is itself of interest, for it was put forth in contention with the idea that crystals grew from a seed or germ:

Germs being inadmissable for explaining the formation of crystals, it is necessary to suppose that the integrant molecules of bodies have each, according to its own nature, a constant and determinate figure, and those molecules which have some analogy between themselves tend reciprocally to approach and unite, sometimes by all their faces indistinctly, sometimes by those faces which are able to achieve the most absolute and immediate contact: but as the *primary elements* of bodies are and probably always will be unknown to us by virtue of the smallness of their parts which escape even the best microscopes, we can only determine the figure of the *secondary elements*.[16]

If Romé de l'Isle had his doubts as to what could be known about molecular form, he was none the less convinced that this form was determined compositely by both the acidic and the basic constituent of the form-giving salt. This was emphasized by him in answer to a criticism which had been leveled at the idea of saline "form-giving" principles. In particular, the Linnaean attempt at correlating saline composition and crystal form had been attacked and the assertion had been made that one of the saline constituents—the acid—seemed to play no role in determining crystal form.[17] Romé de l'Isle defended the Linnaean position by showing that a change in either the acidic or the alkaline constituent would effect a change in the primitive form of the crystal. As justification, he reiterated his view of the dependency of the primitive form on the form of the molecule:

From this observation, it can be concluded that change of form in salts necessarily indicates new modifications in their integrant molecules, although the form of these molecules ordinarily escapes us and can only be surmised.[18]

In the *Essai de cristallographie*, exactly how the primitive form was to be defined for any group of crystals was never made clear by Romé de l'Isle. Eleven years later, he published a revised and greatly expanded edition of this work under the title *Cristallographie*. By 1783 he had made crucial advances towards a quantitative crystallography: he had come upon a measuring device for crystalline angular dimensions—the contact goniometer—and he had been led to enunciate what was to be the fundamental quantitative law of crystallography: the law of constant interfacial angles. This law specified precisely what the formal invariance was which marked crystals of the same substance. According to it, the interfacial angles of crystals of the same substance remained constant and characteristic of that substance no matter how the external forms appeared to differ from one another.[19]

In the interval, Romé de l'Isle had also broadened greatly his conception of the relationship of crystal form to chemical composition and had made this relation more quantitative also. First, under, it seems, the influence of the new chemistry and partly in response to criticisms of his earlier position, he abandoned the idea of a form-giving saline principle. Crystallization now became a characteristic of true chemical union between any substances.[20] Moreover, by chemical composition of a crystal, Romé de l'Isle specified that this meant the *proportions* of the chemical constituents. Thus, the primitive form of a crystal was no longer to be correlated simply with the qualitative constituents of some salt, as was the case in the earlier

---

portions, or with additional principles such as phlogiston. *Ibid.*, pp. 7–8, 34, 37–39.

[14] *Ibid.*, p. 10. Romé de l'Isle himself used the term "cause efficiente."

[15] *Ibid.*, pp. 5, 16, fn. 1 and p. 24, fn. 1 for further references to Macquer's *Dictionnaire* article as well as to Rouelle, whose cubic saline crystals and molecules seem to have been the paradigm for Romé de l'Isle's "primitive form" concept.

[16] *Ibid.*, pp. 13–14. He had in particular the ideas of J. B. Robinet in mind, expressed in his *De la nature* (4 v., Amsterdam, E. van Harrevelt, 1761–1766) 1: pp. 290 ff.

[17] A. F. Cronstedt, *Essai d'une nouvelle minéralogie*, trans. M. Wiedemann, &c.&c. (Paris, P. F. Didot le jeune, 1771), pp. 26–27, 108, 200; J. G. Wallerius, *Minéralogie ou description générale des substances du regne minérale*, trans n.n. (2 v., Paris, J. T. Herissant, 1759) 1: pp. 228–230.

[18] J. B. L. de Romé de l'Isle, *op. cit.*, pp. 20, 15–16, where he suggested that part of the past confusion had been because of a lack of differentiation between the primitive form and the secondary or "accidental" ones.

[19] J. B. L. de Romé de l'Isle, *Cristallographie* (2nd ed., 4 v., Paris, Imprimérie de Monsieur, 1783) 1: pp. xxxiv–xxxv, 70–71. This law apparently was discovered by a pupil of Romé de l'Isle, A. Carangeot, inventor of the contact goniometer. *Cf.* J. Burke, *op. cit.*, pp. 69–71.

[20] J. B. L. de Romé de l'Isle, *Cristallographie* 1: p. vii. The metallurgist, P. C. Grignon, had rebutted Romé de l'Isle's earlier contentions that metallic crystals were not true crystals by citing cases of perfectly regular crystals produced in his own furnace. "Mémoire sur des crystallisations métalliques, pyriteuses et vitreuses artificielles, formés par le moyen du feu," *Mémoires de physique sur l'art de fabriquer le fer* (Paris, Delalain, 1775), pp. 476–481. For the history of work with metallic crystals in the eighteenth century and for the possible influence of Grignon's molecular model of crystals on Romé de l'Isle and Haüy, *cf.* C. S. Smith, *A History of Metallography* (Chicago, University of Chicago Press, 1960), pp. 128–138.

Essai, but now with the proportionate weights of the chemical constituents, whatever they be.[21]

Armed with his law of constant interfacial angles, Romé de l'Isle was able to elaborate somewhat on what he meant by primitive form and how he conceived it to be related to the various external forms actually found in nature. Utilizing an idea of a friend, Romé de l'Isle suggested that these diverse forms could be derived from the common primitive form of a substance if one conceived the edges and/or the solid angles of this latter to be beveled and truncated respectively.[22] But he came up with no more of an articulated model for crystal structure in the *Cristallographie* than he had in the *Essai*. This was not simply due to his lack of ability to theorize; it also was the expression of an innate reluctance to speculate on such matters and to see in Nature the implementation of strict geometry. In the *Cristallographie* he ridiculed the early work of Haüy, which had already appeared, and criticized Bergman as well for what Romé de l'Isle considered to be premature theories of crystal structure based upon the gratuitous destruction by cleavage of a none-too-large supply of crystals.[23]

As a result of his propensities, Romé de l'Isle failed to make more specific how to ascertain primitive forms and relate them to the secondary ones than the still rather vague suggestion of bevelings and truncations. As far as his theorizing on the relationship between the form of the *molécule intégrante* and crystal form, Romé de l'Isle was inconsistent. At one point in the *Cristallographie*, he asserted that the molecules had the same shape as the primitive form;[24] later he denied that one could ever know the shapes of most molecules and at another point, he even implied that the *molécules intégrantes* could vary in size and shape within the same crystal![25]

It remained for Haüy to weld into a closely knit, elegant theory what had been with Romé de l'Isle at best elements in a potential theory of crystal structure. The concept of primitive form, the relationship of primitive to secondary crystal forms, and the function of the *molécule intégrante* as a unit of crystal structure were all worked out with a clarity and coordination which transformed crystallography from the natural history speculation that it still had been with Romé de l'Isle into a mathematical science.

Much of the groundwork for Haüy's synthesis had unquestionably been laid by Romé de l'Isle. But there was an additional source which actually figured with greater prominence in Haüy's first papers on crystal structure and in the *Essai d'une théorie* itself.[26] This was Torbern Bergman. Most famous for his chemical and mineralogical studies, Bergman also had made important studies of crystal structure. Independently of Romé de l'Isle, he had developed a concept of primitive form for certain crystals, notably calcspar and garnet, which was even more precise than Romé de l'Isle's. He had not only postulated primitive forms for these crystalline substances but he had also shown how they could be discovered experimentally by means of crystalline cleavage. If one cleaved a crystal all round symmetrically, the core which remained was the primitive form. Bergman had also given geometrical demonstration of how some of the actual external forms of calcspar and garnet crystals could be generated from their primitive forms. But Bergman had not built into his structural geometry a consistent physical model. The nearest he had come was to suggest that the secondary forms of calcspar crystals were created by a deposit of infinitely thin sheets of crystalline material on each of the faces of the primitive form.[27]

It was thus out of elements brought to light by Romé de l'Isle[28] and Bergman that Haüy fashioned his theory of crystal structure which appeared virtually full blown in the *Essai d'une théorie* of 1784 with little anticipation in his own earlier crystallographical publications. Haüy pictured crystals as being made up of small polyhedral molecules—*molécules constituantes* in the *Essai d'une théorie* but later termed *molécules intégrantes*—stacked together like bricks. These *molécules intégrantes* were of shapes and dimensions specific to each mineral substance. It was the geometry of their shapes and their aggregation which generated the particular forms which crystals of a given substance could assume.

---

[21] J. B. L. de Romé de l'Isle, *Cristallographie* 1: pp. 73–74. *Cf.* below, chap. 2.

[22] *Ibid.*, p. xxvii; J. Demeste, *Lettres du Docteur Demeste au Docteur Bernard sur la chymie, la docimasie, la cristallographie, la lithologie, la minéralogie, & la physique en général* (2 v., Paris, Didot, Ruault, Clousier, 1779) 1: pp. 48–50, 338. Romé de l'Isle was careful to say that this was meant merely as a formal relation, having nothing to do with the way crystals were actually formed.

[23] J. B. L. de Romé de l'Isle, *Cristallographie* 1: pp. xxviii–xxxi.

[24] *Ibid.*, p. 74. His primitive forms were: the tetrahedron, cube, octahedron, rhomboidal parallelepiped rhomboidal octahedron, and dodecahedron with triangular faces.

[25] *Ibid.*, p. 102. The variation in size was to account for truncation, *ibid.*, p. 72.

[26] R. J. Haüy, *Essai d'une théorie sur la structure des cristaux* (Paris, Gogue & Neé de la Rochelle, 1784), pp. 39–41—where his comments are rather snappish. *Cf.* R. Hooykaas, "Les debuts de la théorie crystallographique de R. J. Haüy d'après les documents originaux," *Rev. hist. sci.* 8: (1955) pp. 319–337.

[27] Bergman had used both the terms, "primitive form" and "nucleus." T. O. Bergman, "XII. De Formis Crystallorum Praesertim e Spatho Ortis," *Opuscula Physica et Chemica* (6 v., Upsala, Johan. Edman, 1779–1790) 2 (1780): pp. 2–10.

[28] In fact, nowhere is Romé de l'Isle mentioned in the *Essai d'une théorie*, nor is the law of the constancy of interfacial angles stated there, though Haüy implicitly relied on it. Later, Haüy gave Romé de l'Isle credit for this law. *Cf.* R. J. Haüy, *Tableau comparatif des résultats de la cristallographie et de l'analyse chimique* (Paris, Courcier, 1809), p. xv, fn. 1.

The task of the crystallographer was, as Haüy put it:

> Given a crystal, to determine the precise form of its constituent molecules, their respective arrangement, and the laws which the variations of the [molecular] layers of which it is composed, follow.[29]

Every crystal was envisioned as composed of an inner core, the primitive form or "nucleus" (*noyau*), as Haüy more often termed it, which was revealable by cleavage, and an external secondary form which was the actual shape of the uncleaved crystal. As it had been with Romé de l'Isle and Bergman, Haüy's nucleus was constant and characteristic for the particular crystalline substance. In his theory the secondary form was generated from the nucleus by means of what he called molecular decrements. The nucleus itself was made up of *molécules intégrantes* stacked together contiguously. One was then to imagine a deposition of layers of *molécules intégrantes* on each face of the nucleus in such a way that each successive layer laid down would be recessed, as compared with the one beneath it, by a determinate number of rows of molecules. The result of this kind of molecular deposition would be to build up a molecular pyramid on each face of the nucleus. The coordination of adjacent faces of these pyramids would in turn give rise to a new polyhedron, the secondary form. Molecular recessions or decrements could take place in a variety of ways: each new layer of molecules deposited on a nuclear face could fall short of the layer beneath it by a row (or several rows) of molecules parallel to an edge of the nucleus, or parallel to the diagonal of the nuclear face(thus truncating a corner angle) or even by combinations of these two types of decrements.

One important modification which Haüy subsequently made to his theory had to do with the formal relation between the *molécule intégrante* and the primitive form. In the *Essai d'une théorie*, Haüy thought that they had the same shape in a given crystalline substance. Later he came to believe that there were only three shapes for the molecules: the parallelepiped, the triangular prism, and the tetrahedron, while there were six possible ones for the primitive form.[30] Aside from this, and from a few computational simplifications, his molecular theory of crystal structure remained virtually unchanged.

Although his theory of crystal structure had elements that were arbitrary and circular, it was unquestionably successful in enabling Haüy to relate the primitive form common to crystals of the same substance to the diverse external forms of that substance. The elegance of the model, and its arbitrariness, lay in the assumptions of simplicity which Haüy made about the dimensions of his *molécules intégrantes*[31] and about the rules governing their decrements. An example of how his theory functioned, of which Haüy himself was particularly proud, was his model for iron pyrite crystals. Their external form was a dodecahedron with pentagonal faces. The primitive form was a cube. The *molécules intégrantes* were assumed to be cubic. To generate the external form from the cubic primitive form, Haüy assumed the following rather complex decrements on each face of the cubic nucleus: parallel to one pair of opposite edges, there would be a decrement of one row of molecules for every two molecular layers deposited on the face. On the other pair of edges of the face, there would be a decrement of two rows of molecules for each new layer of molecules. Finally, the adjacent faces of the cubic nucleus were so orientated to each other that the *alternative* decrement took place on contiguous edges. Haüy was able to show by this model of iron pyrite crystal structure not only that the requisite pentagonal dodecahedron was generated from the cubic nucleus but also that (given the assumptions about the molecular form and the laws of decrement) the angle of incidence between two adjacent pentagonal faces was 126°, 52', 8"—close to the measured value of 127°[32] (fig. 1). In all this computation, and this was one of Haüy's more complex models, it is important to note the circumscribing assumptions of simplicity, notably in the cubic form of the *molécule intégrante* and in the numerical ratios of two to one and one to two in comparing rows of decrements to layers of molecules. It was particularly this second prescription that the ratio of molecular decrement to molecular layer be that of small integers that gave Haüy's theory its power, enabling him to predict discrete geometrical relations and angular values, against which the actual measured values could be compared. This element of geometrical discreteness in Haüy's theory of crystal structure was comparable to the gravimetric discreteness which Dalton achieved in the laws of definite multiple proportions in his chemical atomic theory.[33]

Since the *molécule intégrante* was the pivot of his theory, the question faced Haüy of what sort of physical reality to accord to it. He himself was equivocal; he was much too sophisticated a thinker to commit himself to the actual existence of *molécules intégrantes*,

---

[29] R. J. Haüy, *Essai d'une théorie*, p. 25.

[30] R. J. Haüy, *Traité de minéralogie* (1st ed., 5 v., (Paris, Louise, 1801) 1: pp. 28–31. The six primitive forms were: the parallelepiped, octahedron, tetrahedron, hexagonal prism, dodecahedron with rhombic faces, dodecahedron with triangular faces.

[31] For his comments on geometrical simplicity in nature, *cf. Essai d'une théorie*, pp. 25 ff. and *Observations sur la mésure des angles des cristaux* (Paris, Huzard, 1818).

[32] R. J. Haüy, *Traité de minéralogie* (1st ed.) 1: pp. 39–41; 5 (atlas): figs. 14–16.

[33] This idea was suggested by C. Mauguin. "La structure des cristaux d'après Haüy," *Bull. Soc. fran. Minéral.* 67 (1955): pp. 244–245.

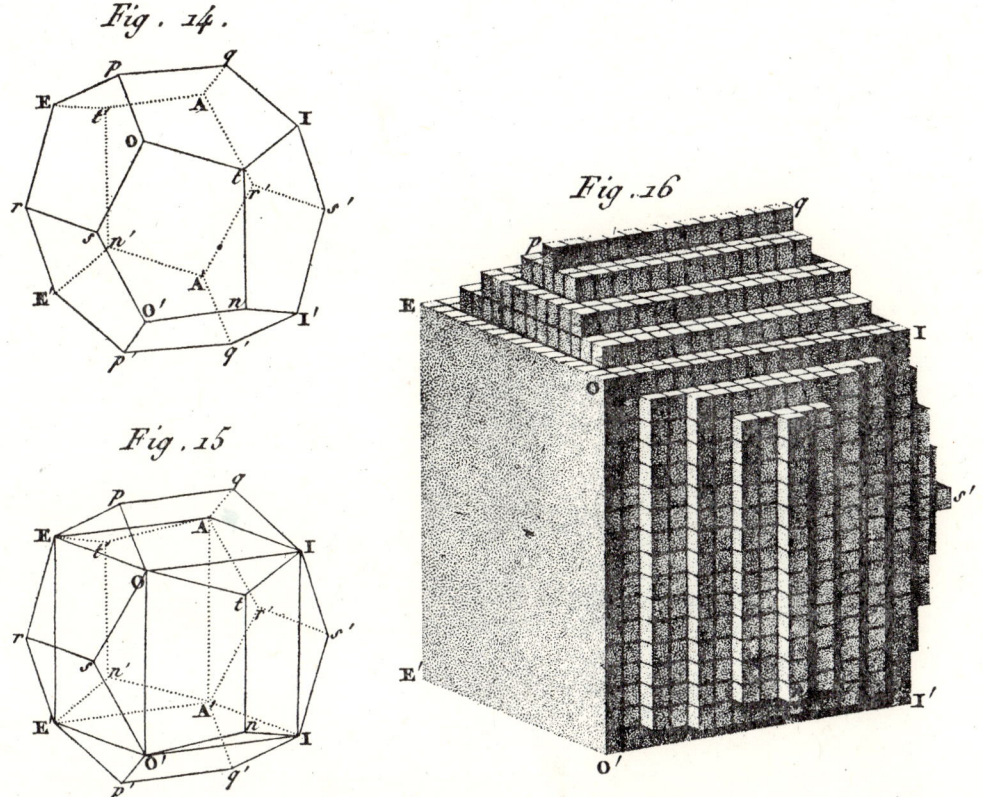

FIG. 1. Haüy's model of the molecular structure of iron pyrite. *Traité de minéralogie* (1801).

or to the physical reality of the structural model based upon it. Although he seemed to accept such reality in the *Essai* of 1784, in later works—the *Traité de minéralogie* of 1801 and subsequent publications—Haüy was always careful to qualify any statement which appeared to posit the existence of the *molécules intégrantes*.[34]

In fact there were good physical reasons not to make such a commitment. For if the power and elegance of Haüy's molecular model came from the molecules being conceived as contiguous to each other, that is, establishing material continuity, there were reasons why such continuity seemed unlikely. First, there were cases where the actual structural model seemed to necessitate the interposition of vacua among the molecules, the most prominent example of this, the case of fluorspar, being known to Haüy from the inception of his model. This case continued to plague him and to provide ammunition for the critics of his theory.[35] More serious was an argument from optics, a field in which Haüy himself carried out research. Here it had to do with the need to explain transparency in crystals. In the Newtonian theory of the propagation of light, accepted by Haüy and his contemporaries, transparency was thought to be the effect of light corpuscles passing through a medium without encountering any material molecules to deflect them. Transparent crystals, then, like all other transparent media, had to be virtually empty of matter.[36]

These physical arguments were strong ones against the "building-block" model of crystal structure, and Haüy was aware of them. Nevertheless, though he scrupled against asserting the physical reality of his model, in practice he seems to have regarded his *molécules intégrantes* as the actual structural components of crystals, rather than as merely mathematical constructions. The domain of his interest in which this was most telling was in mineralogy, where Haüy came to identify the *molécule intégrante*, not only as the structural unit of crystals but also as the compositional

---

[34] R. J. Haüy, *Traité de minéralogie* (1st ed.) **1**: p. 21.

[35] *Cf.* W. H. Wollaston, "The Bakerian Lecture. On the Elementary Particles of Certain Crystals," *Phil. Trans.* **103** (1813): pp. 51–53; J. J. Bernhardi, "Gedanken über Krystallogenie und Anordnung der Mineralien," *Journal für die Chemie, Physik und Mineralogie* (Gehlen's Journal) **8** (1809): pp. 367–371; H. J. Brooke. *A Familiar Introduction to Crystallographie* (London, W. Phillips, G. Yard, 1823), pp. 42, 65–66.

[36] R. J. Haüy, *Tableau comparatif*, p. 126. Another argument against his model of contiguous molecular building-blocks, whose implications he never seemed to notice however, was implicit in his mineral species concept, in which the non-essential constituents were considered as simply mechanically interposed between the "essential" molecules. Below, chap. 2.

unit of all minerals—as the chemical molecule of John Murray's definition.

Haüy himself was not particularly interested in the details of how the *molécule intégrante* was formed out of its constituents; it was not the *molécule intégrante* as the product of chemical reaction with which he was concerned. Nor was he even much interested in speculating about the shapes and arrangements of the elementary molecules in the *molécule intégrante*. The few cases where Haüy compared the shapes of the constituent crystalline molecules with that of the compound (this was only feasible in a very limited number of cases) were disappointingly inconclusive.[37]

But there is no question that Haüy himself believed that his *molécules intégrantes* were formed by chemical association of elementary molecules of diverse natures and determinate shape; that they then agglomerated in orderly fashion in crystals; that the form of the *molécule intégrante* was invariable for that chemical substance; and that this form was correlative with the molecular composition, also invariable for the substance. This came out most clearly when Haüy turned his attention to mineralogy. Here, he, along with Romé de l'Isle, and the contemporary geologist, Deodat Dolomieu, enunciated a doctrine of fixed mineral species, defined by crystallographical and chemical criteria. For Dolomieu and Haüy in particular, the key to their mineral species was the *molécule intégrante*. In this mineralogical doctrine the implication of a molecular concept of matter as having definite and fixed composition was first realized.

II. THE MOLECULAR "INDIVIDUAL"

When Haüy turned his attention to mineralogy in the 1790's, the classification of rocks and stones was on its way to becoming a science, but it had not yet reached scientific maturity. Mineral classification, almost as ancient as its two sisters of the organic kingdom, zoology and botany, had effloresced vigorously in the eighteenth century, supported by the amateur's curiosity, the miner's and metallurgist's practical need, and the naturalist's desire for order and systematization.

A number of different approaches to classifying minerals had been developed. By the last quarter of the eighteenth century, three distinct types of systems were discernible: those based upon chemical analysis, those based upon a comparison of external characteristics of minerals, and those which emphasized crystal form. In the first approach, the simplest in terms of its assumptions if perhaps the most difficult to put into practice, minerals were classified according to their chemical nature. This was perhaps the most widespread method of mineral classification in the eighteenth century and continued to be very prominent throughout the nineteenth. Its foremost eighteenth-century proponents included A. F. Cronstedt, Torbern Bergman, and Richard Kirwan.[1] The second method, which might be called the "naturalist's," was developed most systematically by the German geologist and mineralogist, A. G. Werner and his followers at the great mining school at Freiberg of which Werner was head.[2] Werner and his disciples believed that, for practical purposes (ease of identification) as well as for the accuracy of the classification, it was essential to know the readily identifiable external characteristics of minerals, their color, texture, hardness, and shape. The miner and the naturalist would only be truly served when they had at hand systematic lists of these characteristics by which they could accurately identify their minerals. The third type of classification, based on crystal form, was developed particularly by Romé de l'Isle, Haüy, and Dolomieu.

None of these types of classification systems were exclusive; in particular, all recognized the ultimate desideratum of chemical composition in any mineral classification scheme. The question involved in the two non-chemical approaches was whether there were characteristics of minerals other than the chemical composition as discovered by chemical analysis which revealed the "essential" similarities and differences between minerals. It was particularly the third approach, the crystallographical, which claimed the ability to do this.

Connected with this claim was another issue, this one of even broader scope: the whole question of the principles upon which mineral classification were based. In this period when mineralogy was developing under the impetus of practical needs and the naturalists' interests, one might have expected that considerable attention would have been given to the theoretical problems of mineral classification. Was it possible to classify

---

[37] He attempted to determine whether his "antimonial silver" was a chemical compound or an alloy. In the latter case, the crystalline form ought to reflect that of either silver or antimony. *Traité de minéralogie* (2nd ed., 5 v., Paris, Bachelier, 1822) **3**: pp. 263–264. His inability to perceive how molecules of silver and of antimony could unite to produce a compound molecule which was formally incompatible with *both* led him to speculate very tentatively on chemical elements of a lower order than the accepted ones.

[1] A detailed study of pre-eighteenth- and eighteenth-century mineralogy has recently been completed by David Oldroyd: *From Paracelsus to Haüy: The Development of Mineralogy in its Relation to Chemistry* (Ph.D. Dissertation: University of New South Wales, 1974), *Ambix* **21** (1974): pp. 157–178. *Cf.* also, R. Hooykaas, "The Species Concept in 18th Century Mineralogy," *Arch. int. hist. sci.* **5** (1952): pp. 45–55. Much of the material in this chapter has been published in my article, "Minerals, Molecules and Species," *ibid.* **23** (1970): pp. 185–206.

[2] A. G. Werner, *Von den äusserlichen Kennzeichen der Fossilien* (Leipzig, S. L. Crusuis, 1774). An English translation by A. V. Carozzi: *On the External Characters of Minerals* (Urbana, University of Illinois Press, 1962).

minerals? If so, how did such classification parallel that of the biological kingdoms? What was the best method of ascertaining mineralogical division: was it through elaborate descriptive categories, or did some of the developing physical sciences offer simpler, more elegant, and "truer" criteria? Such questions were raised, but not generally. Rather, it seems that the locus for such discussions became the above-mentioned group of French mineralogists. In particular, their considerations focused upon the question of how to define mineral species, the fundamental taxonomic unit.

This group attempted to provide mineralogy with a firm theoretical foundation of its own; they all strongly asserted that fixed and distinct mineral species existed.[3] In justifying their position, they were led to take up the question of how closely and in what ways mineral taxonomy paralleled classification in the organic kingdom. These questions became particularly important in the latter half of the eighteenth century. For by this time a clear-cut demarcation between organisms on the one hand and minerals on the other was emerging, although a few writers asserted that minerals "grew" like organisms or, conversely, that organisms developed like minerals.[4] In particular, these mineralogists faced the questions of what corresponded among minerals to the concept of "individual" and "species" in the organic kingdom.

Mineral classification had figured in the great taxonomic surveys of Linnaeus and Buffon, published throughout the middle of the eighteenth century, and, although neither of them explicitly grappled with types of questions raised above, their approaches to mineral classification must be reviewed because of their influences on the French mineralogists. Implicitly divergent in a number of ways at least, Linnaeus's and Buffon's points of view served as something of a dialectical context for the debate over mineral species; Linnaeus's being the guide for the French mineralogists, Buffon's providing the critique which they had to answer.

For his organic classification Linnaeus had assumed that there existed a constant number of species. He had never clearly distinguished living entities from non-living ones and in his classificatory scheme he made use of parallel assumptions. Just as he had achieved striking success in botany by the use of botanical forms, so he attempted to classify minerals according to crystal form—a position reinforced by his belief that minerals were formed by some kind of "impregnation" of water by a form-giving "salt" to form a distinctly shaped crystal. In his implicit belief in fixed mineral species and his use of crystal forms for classification, Linnaeus was the progenitor of the French mineralogists to be discussed.[5]

In the *Premier discours* of the *Histoire naturelle*, Buffon had opened his great survey of Nature with a sharp criticism of Linnaeus's entire taxonomy for having imposed disjunction upon a Nature really marked by continuity. To Buffon there were no sharp divisions in Nature; any system of classification into distinct species, genera, etc., reflected therefore not what really existed in Nature, where insensible gradation producd a continuum through the entire length of the *scala naturae*, but rather categories of the mind—and categories which resulted in a false apprehension of Nature.[6] Buffon subsequently modified his stand. He finally accepted the idea that the organic kingdoms were, in fact, divided into distinct species.

Buffon had not explicitly compared classification[7] in the mineral kingdom with that of the organic kingdoms any more than Linnaeus had done. But, though Buffon never envisaged a complete disjunction between organisms and minerals, he did see essential differences between them; organisms had characteristics which minerals did not possess. Moreover, it was these peculiar organic characteristics which Buffon invoked when he arrived at a concept of "fixed organic species." They were the existence of an interior organization, and the ability to perpetuate this organization through reproduction.[8]

The entity by which Buffon saw the perpetuation of uniform species in organisms was the *molécule organique*. Interestingly enough, the origins of Buffon's concept of the organic molecule seem to have been in crystal structure studies, particularly those of G. F. Rouelle. Rouelle's studies of common salt crystals showed that these crystals, cubic in shape, seemed to be made up of little cubic elements—and, by extrapolation, of cubic corpuscles. Reasoning analogously from crystal formation to the development of organisms, Buffon postulated an organic molecule which conditioned the developmental form of organisms. There were important differences of which Buffon was well aware, the most difficult one being that organisms developed highly complex and differentiated internal structures. Buffon brushed aside the obvious problems of what the shapes of these organic molecules were in relation to the forms and structures of their organisms, and exactly how they

---

[3] There are a few works which deal with this doctrine of fixed mineral species, but none has placed it in the biological context in which it arose. *Cf.* J. Orcel, "Haüy et la notion d'espèce en minéralogie," *Bull. Soc. fran. Min.*, **67** (1944): 265–337; R. Hooykaas, "The Concept of the 'Individual' and 'Species' in Chemistry," *Centaurus* **5** (1956–1958): pp. 307–322.

[4] H. Metzger, *La génèse de la science des cristaux*, pp. 93–123. *Cf.* also J. B. L. de Romé de l'Isle, *Cristallographie* **1**: pp. 17–22.

[5] Above, chap. 1, footnote 11.

[6] G. L. LeClerc, Comte de Buffon, "Premier discours," *op. cit.* **1** (1749): pp. 13 ff.

[7] *Cf.* "De la reproduction en général," *ibid.* **2** (1749): p. 18. Much more emphatically, "De la nature, seconde vue," *ibid.* **13**: pp. i–x.

[8] "Comparaison des animaux, des végétaux & des minéraux," *ibid.* **2**: pp. 1–17.

determined the macrostructure of the organisms. To him, the illumination of the analogy overrode these vexing questions.[9]

By whatever mechanism and means, the organic molecules were Buffon's carriers of the species from generation to generation—molecules, passed on from parents to children which contained somehow, in microcosm, the organization of the species and insured that the progeny would develop the same kind of interior organization that the parent had possessed. It was in these *molécules organiques* that the sum of living matter was contained. They had themselves been shaped out of brute matter by the interaction of the expansive force of heat and the attractive force of gravity. Brute matter in itself was endowed with no organization.

However, it sometimes received at least some organization in crystal structure. One type of model for crystal structure, that of the accretion of corpuscles of the same shape as the crystal, influenced his concepts on organic growth and development. Accepting this, Buffon suggested an obscure mechanism for what he called the efficient caused of crystalline form: that crystalline structure was somehow brought about by the presence of organic molecules in a material mass. But while the organic molecules determined the internal structures of organisms, they could only act superficially in brute matter—"in two dimensions rather than three." Therefore, crystals increased by the accretion of mineral layers to the external surface rather than by the articulation of an internal structure. Moreover, even crystals lacked the power of reproduction. By implication at least, species, as Buffon came to define them, could not exist in the mineral kingdom.[10]

With regard to crystal forms, Buffon denied that there existed any invariance which might lend them to use in mineral classification. In his late work on minerals, published in the 1780's, he derided *"nos cristallographes"* who had attempted to relate the various forms of calcspar to each other by models of superposed rhombohedral molecules as having "merely substituted ideal combinations for the real facts of nature."[11] Noting that there were cases of different substances crystallizing in the same form and the same substance crystallizing in a variety of apparently unrelated forms, he concluded that crystalline form, far from being constant, was "the most equivocal and variable of all the characters by which minerals ought to be distinguished."[12]

But neither Linnaeus nor Buffon had explicitly dealt with the general principles of mineral taxonomy or compared the method of classifying minerals with those used in the organic kingdoms. Rather, these questions seem first to have been raised by Buffon's erstwhile collaborator at the Jardin du Roi, Louis Daubenton. He did so in the Introduction to the first volume of his *Histoire naturelle des animaux* (1782) in the *Encyclopédie méthodique*.

Daubenton was a wide-ranging naturalist. Primarily an anatomist and zoologist, he also took a keen interest in mineralogy. He taught it at the Jardin du Roi and he held the first chair in mineralogy in the Muséum d'Histoire Naturelle, the revolutionary reconstruction of the Jardin, from 1793 until his death in 1799.[13] With his range of naturalistic interests, Daubenton was in a good position to appreciate the questions of how classification of the mineral kingdom compared with that of the organic kingdoms and of whether any system of mineral classification reflected reality in nature.

Like his mentor Buffon, Daubenton was skeptical of apprehending reality in nature through the divisions of classification. But he approached this problem, as it were, from the opposite direction to Buffon. Originally at least, Buffon had seen continuity in nature as its fundamental quality. But to Daubenton, it was discreteness by which nature was characterized: "There are really and distinctly only individuals among plants and animals."[14]

Nevertheless, it was precisely this fact which enabled taxonomists to classify them. It was the individuals which could be compared with each other and arranged into species on the basis of their similarities and differences. The species thus depended upon the individuals. They were real because of the reality of the individuals which composed them. They were permanent and invariable because of the succession of individuals through reproduction.[15] However, the species was the only taxonomic division which possessed some

---

[9] "De la reproduction en général," *ibid.*, pp. 18–41, especially pp. 23 ff. *Cf.* J. Roger, *Les sciences de la vie dans la pensée française au XVIIIᵉ siècle* (Paris, Armand Colin, 1963), pp. 542–558, for the general context of Buffon's ideas on generation. *Cf.* also, H. Metzger, *La génèse de la science des cristaux*, p. 105 ff.

[10] G. L. LeClerc, Comte de Buffon, "De la figuration des minéraux," *op. cit.* 2: pp. 1–17.

[11] "De la pierre calcaire," *Histoire naturelle des minéraux* (5 v., Paris, Imprimerie royale, 1783–1788), 1: pp. 241–242. He probably meant Haüy here.

[12] "Du plâtre et du gypse," *ibid.*, p. 343. To this, Romé de l'Isle answered in his *Cristallographie:* "Je sens combien est imposante l'autorité d'un homme tel que M. le Comte de Buffon, mais il produit son Ouvrage, je produis le mien: c'est aux Naturalistes à decider de quel côté se trouve l'erreur ou la vérité," p. xvii, footnote 2. Even more emphatic against the "Cristallographes" was Buffon in "Des Cristallisations," *Histoire naturelle des minéraux* 3: pp. 425–434, especially 433, where he wrote against them that "dans la Nature, il n'y a rien d'absolu, rien de parfaitement régulier."

[13] For biography *cf.* the article "Daubenton," in the *Nouvelle biographie générale,* ed. F. Hoefer (46 v., Paris, Fermin Didot frères, 1855–1866) 13: pp. 162–166.

[14] L. Daubenton, "Introduction à l'histoire naturelle," *Encyclopédie méthodique. Histoire naturelle des animaux* (4 v., Paris, Panckoucke, 1782–1789) 1 (1782): p. iii.

[15] *Ibid.*

status of reality for Buffon. He denied that there was any *scala naturae* linking all the productions of nature into one grand succession; therefore all divisions higher than species were merely arbitrary creations of the taxonomist.[16]

Daubenton also considered the classification of minerals. Here he saw a sharp distinction between the organic kingdoms and the mineral: minerals differed from plants and animals because they lacked organization and life.[17] These two differences had serious consequences for the classifiers of minerals. Since minerals possessed no internal organization, it was impossible to specify what was the mineral analogue to the "individual" organism. In the absence of individuals there could be no species, since the species was inseparable from the individuals subsumed in it. Moreover, since there were no mineral individuals, there was no reproduction in any usual sense of the word. There was, therefore, no way of passing on a constant set of specific characteristics from one generation to another.[18]

Daubenton seems to have played a pivotal role in turning Romé de l'Isle and probably Haüy to consider the theoretical basis of mineralogy. Romé de l'Isle explicitly responded to Daubenton's assertions. Haüy attended Daubenton's lectures on mineralogy at the Jardin du Roi and his early writings on mineralogy reflected Daubenton's views and language. When he turned away from these views, he did so with particular circumspection.[19]

It was to rebut Daubenton that Romé de l'Isle took up the subject of mineral taxonomy in a work entitled *Des caractères extérieurs des minéraux* of 1784.[20] By this time, he had already published his two important works on crystallography: the *Essai* of 1772 and the expanded version of it, the *Cristallographie* of 1783. All his crystallographical work had implied the existence of fixed and distinct mineral species,[21] but it was first in the *Des caractères extérieurs* that he actually argued this thesis in order to answer Daubenton's assertion that mineral species did not exist:

Thus, this mineral kingdom, this assemblage of bodies called brute inorganic because they are deprived of all interior apparatus of organs necessary to life, to growth, to reproduction, this mineral kingdom, I say, has SPECIES in particular, as constant, as determinate, according to the invariable laws of combination, of saturation, as the animal and plant species themselves are, according to laws no less certain, of fecundation.[22]

As this makes evident, Romé de l'Isle took constant chemical composition as the defining characteristic of his fixed mineral species. Invariability of the laws of chemical affinity guaranteed just the same fixity of mineral species over time as did reproduction for the organic species.[23]

Curiously enough, even though he was explicitly challenging Daubenton's denial of the existence of mineral species, Romé de l'Isle failed to answer Daubenton's main point that there were no mineralogical individuals. It was just this question with which Haüy and, even more emphatically, Dolomieu were to grapple and to suggest an answer centering upon the *molécule intégrante*. But, just as Romé de l'Isle had not incorporated the *molécule intégrante* into a consistent model of crystal structure, so he evaded dealing directly with Daubenton's point. One might infer from his discussion of mineral species that he had in mind a mineralogical "individual"—perhaps "homogeneous" mineral samples or possibly even *molécules intégrantes*—but he failed to make either of these possibilities explicit.

---

[16] *Ibid.*

[17] *Ibid. Cf.* also, "Les trois règnes de la nature," *ibid.*, pp. xi–xii (in the context of vegetable debris, fossils, etc. no longer being part of the vegetable kingdom).

[18] "Parmi les minéraux, il n'y a point d'individus & par consequent point d'espèces. Nous ne voyons pas que les minéraux se reproduisent comme les plantes & les animaux, par des individus semblables, de génération en génération. Un minéral s'altère & se détrait par divers accidens; ses parties intégrantes se dispersent, se mêlent & se combinent avec des minéraux d'autres sortes, souvent très-différens de celui qui a été décomposé. Il n'y a point là d'individus, puisqu'il n'y a point de ressemblance essentielle." *Ibid.*, p. iii. *Cf.* also, *ibid.*, p. xii.

[19] Haüy and Daubenton were close friends and colleagues up to the latter's death in 1799. *Cf.* A Lacroix, "Le vie et l'œuvre de l'abbé René Just Haüy," *Bull. Soc. fran. Min.* 67 (1944): p. 21 ff. It was Daubenton who encouraged Haüy to preesnt his theory of crystal structure to the *Académie des Sciences. Cf.* R. J. Haüy, *Essai d'une théorie* . . . , pp. 38, 5–6, where he spoke highly of Daubenton's mineralogy, significantly here, in context of the at best subsidiary role of crystal form in mineral classification.

Daubenton was also close with Dolomieu; he had given him advice and information about Sicily prior to Dolomieu's geological exploration there. *Cf.* A. Lacroix, *Deodat Dolomieu* (2 v., Paris, Perrin, 1921) 1 pp: xvii, 79 and letter VI (80–83).

[20] "Un Professeur d'Histoire Naturelle, justement célèbre par ses profondes connoissances en anatomie, répète tous les jours à ses élèves, 'qu'il n'y POINT D'INDIVIDUS, et par consequent POINT D'ESPÈCES parmi les minéraux mais seulement des variétés dont la collection peut composer différentes SORTES de minéraux.' Ces assertions ont déjà passé dans quelques traités élémentaires de minéralogie & meritoient d'autant mieux d'être approfondies . . ." J. B. L. de Romé de l'Isle, *Des caractères extérieurs des minéraux* (Paris, the author, 1784), p. 1. *Cf.* above, footnote 18.

[21] *Cf.*, for example, his *Cristallographie*, xxxiv–xxxv, where he implied constancy of mineral species from the law of constancy of interfacial angles, this being his first mention of this law.

[22] J. B. L. de Romé de l'Isle, *Des caractères extérieurs des minéraux*, pp. 55–56.

[23] His actual definition of mineral species was as follows: "Les corps homogènes de la même espèce seront donc ceux *qui admettront dans leur composition non seulement les mêmes principes constituans, mais encore une quantité déterminée de ces mêmes principes. Ibid.*, pp. 9–10. Romé de l'Isle appealed to the idea of saturation of chemical affinity to justify his assertion of fixed mineral species. *Ibid.*, p. 56, footnote 1. *Cf.* above, chap. 1, footnote 21.

Moreover, he fulminated against the use of chemical analysis in mineralogy even though his definition of mineral species was a chemical one. He was loath to see mineralogy, to him the province of the naturalist, become merely a branch of chemistry and he did not think that chemical analysis was sufficiently perfected to be used in classifying minerals into their species.[24] Instead, he preferred certain physical characteristics such as crystal form, density, and hardness as more reliable attributes for mineral taxonomy. He selected these because he believed that they reflected chemical homogeneity with particular directness and they persisted to indicate the "essential" chemical constituents of mineral samples even when these were "contaminated" by foreign admixtures.[25]

The year in which Romé de l'Isle published his *Des caractères extérieurs des minéraux* was the same as that in which Haüy's epochal *Essai d'une théorie sur la structure des cristaux* appeared. Although Haüy laid down in this work the theory of crystal structure which remained virtually unchanged during the rest of his career, he was slow to turn his attention to mineralogy and to bring to bear on it his theory. Indeed, as he himself admitted, he originally saw little relevance in his theory of crystal structure to mineral classification[26]—a strange oversight if one accepts Cuvier's word, for, according to that biologist, Haüy was originally attracted to crystallography from botany by his desire to find morphological constancies among minerals analogous to those with which the botanist dealt among plants.[27]

But, after a slow start, Haüy did turn his attention to mineralogy and he gradually evolved a system based upon principles similar to those of Romé de l'Isle, but which differed from that of his predecessor in two important respects. First, Haüy was much more directly concerned with relating chemistry and crystallography to mineralogy, and to each other. Second, the locus of these relationships was placed by Haüy in the *molécule intégrante*.

In his first published work on mineral classification, Haüy had not yet come this far. The paper, a treatise on mineralogical methods published in 1793, dealt with the problem of the existence of mineral species. He introduced the discussion of this question by an account of the process involved in naming things. His approach here, similar to that of Condillac, was that nature could be clarified and understood if only we worked out an analytical language to describe it. Haüy took care to assert that at least the lowest division of classification, the species, existed not only in the mind of the classifier, but in nature as well. This held true not only for organic species, but for mineral ones too. These latter were defined in terms of constant chemical composition:

If we now examine the basis of the different divisions and subdivisions of the method [of classification] we see first that the species is given immediately by nature. In botany, for example, it is the reproduction of individuals, one by another, which properly determines the species; and it is only in consequence of that reproduction that all individuals of the same species are similar in all their parts. ... In mineralogy, the fragments of the same species are those which have the same principles combined together in the same proportions.[28]

Like Romé de l'Isle's definition, Haüy's left a number of fundamentals unanswered, the most obvious of these being whether there was anything among minerals which corresponded to the individual organism. However, unlike Romé de l'Isle, Haüy did not ignore this question, or the implication Daubenton had drawn from it. Indeed, he took up Daubenton's stricture against mineralogical "individuals" and accepted Daubenton's position:

I employ here the word, species, in imitation of most naturalists, instead of substituting the word, *sort,* as have MM. Buffon and Daubenton, because, after all, it is only a matter of determining the sense which one attaches to this word when one speaks of a mineral. But the term, *individual,* seems to me to be too significative to be applied to a mineral, each of whose parts is always the same mineral, whereas an animal or a plant cannot be divided without losing the character which constitutes it as an individual.[29]

Thus, Haüy had defined species in terms of fixed chemical proportions for minerals in order to get around the impasse raised by Daubenton. But this only raised a further question: how was the mineralogist to deal with minerals of variable composition? What were the criteria by which the "essential" chemical constituents of a mineral species could be determined in a given mineral sample? Haüy was not yet ready to grapple with these questions and he wrote rather lamely in the 1793 treatise that his method was intended primarily

---

[24] *Ibid.*, pp. 45, 49–50 for examples.

[25] *Ibid.*, p. 4. Of these three characteristics, Romé de l'Isle wrote: "... il n'existe point dans la Nature deux substances intrinsiquement différentes, qui aient en même temps la même forme cristalline, la même pesanteur & la même dureté spécifiques." *Ibid.*, p. 12. He did relate these characteristics, and especially crystal form, which had pride of place, to the nature of the mineral *molécule intégrante*.

[26] R. J. Haüy, *Essai d'une théorie* ... , pp. 4–6. *Cf.* above, footnote 19. In the *Tableau comparatif*, pp. xxxii–xxxiii, Haüy admitted that he had come to change his mind over the importance of his crystal structure theory to mineral classification.

[27] G. Cuvier, "Éloge historique de M. Haüy," *Paris, Mém. Acad. Sci.* 1825, **8** (1829) : p. cl.

[28] R. J. Haüy, "Mémoire sur les méthodes minéralogiques," *Annal. de chimie,* **18** (1793) : pp. 230–231. Haüy supported the position here that chemical analysis was the fundamental criterion for mineral classification. *Ibid.*, pp. 231, 232, footnote a. External characteristics, particularly formal ones, were "... à la distinction des modifications & des variétés ..." *Ibid.*, p. 237, footnote a.

[29] *Ibid.*, p. 231, footnote a.

"for substances in their greatest state of purity; because they are the substances which represent properly the species." [30]

By the time Haüy made his next comprehensive statement of his mineralogical ideas, he had obviously reconsidered and resolved the question of the mineralogical individual, and he had also confronted the issue of chemical purity. Both these issues were taken up in his monumental *Traité de minéralogie* of 1801 and he approached them by focusing upon the unit of his crystal structure theory, the *molécule intégrante*.[31]

With regard to the question of the mineralogical individual, Haüy had in 1793 denied that this term could be applied to any mineralogical entity expressly because there seemed to be nothing among minerals which corresponded to the individual organisms—entities defined as the smallest units of a species which retain the specific characteristics. But Haüy himself had already defined the *molécule intégrante* in precisely these terms.[32] In the *Traité de minéralogie*, Haüy employed his molecular individual as such in all but name. The *molécule intégrante* became the basis of his taxonomy and its geometric essence, observable in crystal form and structure, the basis of his method. What seems to have provided the impetus for Haüy to modify and sharpen his ideas on mineral taxonomy was the difficulty in dealing with discrepancies between the expected correlations of crystallographical and chemical data. Even though he had not relied on crystal form in his earlier mineralogical scheme, from the first there had been implicit in his molecular crystal structure theory a strict correlation between crystal form and chemical composition. For most minerals there had been excellent agreement in this respect; this was what had given Haüy confidence in a taxonomic system based upon chemical composition in the first place. But matters had not remained so simple. Discrepancies—particularly variability in chemical composition of minerals which appeared to Haüy to belong to the same species on the basis of their crystal form—caused Haüy to re-evaluate the roles of chemical analysis and crystallographical determination in mineral classification.

It was not that Haüy lost faith in his belief that fixed mineral species, characterized by constant chemical composition, existed. Rather, he came to mistrust the ability of chemical analysis to get at the true mineral species. In his new view, chemical analysis was incapable, in itself, of separating the unvarying "essential" components of minerals from those which were foreign to the species and were merely "interposed." Instead, crystal form became the best criterion, for it reflected the constant specific characteristic of what now became the locus of the species: the *molécule intégrante*:

> There exists a characteristic much more solid and proper, by its invariability, to serve as the rallying-point for different bodies which belong to the same species. This is that which derives from the exact form of the integrant molecule, because that form subsists, without any sensible alteration, independently of all the causes which can make the other characteristics vary.[33]

Two constant specific characteristics, crystal form and chemical composition, with the former dominant over the latter in terms of actually determining species, came to define mineral species for Haüy and they were designated as the fundamental properties of the *molécules intégrantes*. Haüy's definition of mineral species showed the central position the *molécule intégrante* had now been given in his mineralogical considerations, and its implicit role as the mineralogical individual:

> ... the species, in mineralogy, *a collcetion of bodies whose integrant molecules are similar and which are composed of the same elements united in the same proportion*.[34]

But the ideas which Romé de l'Isle and Haüy had been developing received their most "philosophic" and consistent treatment from a naturalist outside the crystallographical tradition. This was D. Dolomieu, geologist, mineralogist, and adventurer. The work in which Dolomieu expounded these ideas was his *Sur la philosophie minéralogique et sur l'espèce minéralogique*, published by him in 1801. This work was written in difficult circumstances, to say the least. Dolomieu had accompanied Napoleon on the Egyptian expedition of 1798. On his return, in Sicily he fell into the hands of an old enemy in the Order of the Knights of Malta. He was imprisoned in Messina under dreadful circumstances until Napoleon secured his release in 1801. It was during this imprisonment that he wrote his *Philosophie minéralogique*, which was published on his return to Paris.[35]

---

[30] *Ibid.*, p. 239.

[31] The development of Haüy's ideas on species classification is somewhat confused by the fact that already in 1792, he had asserted that crystal structure (as opposed merely to crystal form) was "un point fixe et invariable, relativement à tous les corps d'une meme espèce" and, "sans mener aussi loin que l'analyse chimique," it ought to play an important role in species determination. R. J. Haüy, "De la structure considérée comme caractère distinctif en minéralogie," *Jour. d'hist. nat.* 2 (1792): pp. 56–71 (70 and 56 especially, for quotations).

[32] *Cf.* R. J. Haüy, "Exposition de la théorie sur la structure des cristaux," *Annal. de chimie* 17 (1793): p. 253.

[33] R. J. Haüy, *Traité de minéralogie* 1 (1801): p. 156. He dismissed the objection that well-formed crystals were rare by asserting ". . . si en objectant que les cristaux sont rares, on veut dire qu'il a beaucoup d'espèces minéralogiques qui ne se présentent jamais sous des formes cristallines, je demanderai si ce sont proprement des espèces, et non pas plutôt des mixtes à la production desquels différentes espèces ont concouri." *Ibid.*, p. 159, footnote 1.

[34] *Ibid.*, p. 162.

[35] A. Lacroix, *Déodat Dolomieu*, 1: pp. xv–xliv for life. *Cf.* also K. L. Taylor, "Dolomieu, Dieudonné (called Déodat) de Gratet de," *Dictionary of Scientific Biography* 4: pp. 149–153.

Although Dolomieu was not an innovator in crystallography, he was an important geologist and mineralogist—a friend of Daubenton and Haüy.[36] On Daubenton's death in 1799, Dolomieu was elected *in absentia* to succeed to the chair of mineralogy at the Muséum. Dolomieu in turn was followed by Haüy in 1802. Dolomieu was thus unquestionably familiar with both the questions of mineralogical methodology raised by Daubenton and with the advances in the theory of crystal structure made by Romé de l'Isle and Haüy. His *Philosophie minéralogique* represents an integration of Haüy's (and perhaps Romé de l'Isle's) theory of molecular structure with Daubenton's comparative analysis of mineralogical method.[37] In a synthesis which Haüy approached but did not achieve in his own *Traité de minéralogie,* also of 1801, Dolomieu gave direct answer to Daubenton's objection to the existence of mineral species by focusing on the *molécule intégrante.*

Dolomieu's position in the *Philosophie minéralogique* was very close to that of Haüy in the *Traité de minéralogie.* Like Haüy, he viewed the *molécule intégrante* as the seat of mineral species. But his position was not precisely that of Haüy. He was less concerned than the crystallographer (in the *Traité*) with asserting the *priority* of crystallography over other sciences in determining mineral species. Beyond that, Dolomieu was also less interested than Haüy in practical classification and more concerned with establishing the philosophical basis for mineralogy. In line with this broader approach of Dolomieu, he went beyond Haüy to locate mineralogical "individual" in the *molécule intégrante:*

> By the word, *individual,* is meant an object which cannot be divided without losing its complete existence, without ceasing to be itself. But the shapeless mass of an homogeneous mineral can be divided and subdivided without each part, separated from it, becoming essentially different from the whole; it loses its volume without the species losing its existence, and that subdivision can proceed down to the integrant molecule. But beyond that limit, where mineralogy stops and chemistry begins, any new division destroys the individual, in separating its constituent principles, in isolating its elements from its institution.[38]

Not only did the *molécule intégrante* guarantee the existence of mineral species for Dolomieu, but, by virtue of its formal and compositional invariance, it assured the existence of fixed, definite mineral species, between which "no imperceptible nuances, no insensible gradations" were to be found.[39] Thus, the objections raised by Daubenton to the existence of mineral species were finally fully answered by Dolomieu: the mineralogical individual was found in the theoretical structural and chemical unit, the *molécule intégrante.* And, like Romé de l'Isle and Haüy, Dolomieu saw as the exemplar of the mineral species, amenable to our senses, the regularly shaped, chemically homogeneous crystal composed of *molécules intégrantes.*

Although the lines of influence and connection between them are not as clear as one might wish, it is apparent that these three mineralogists, Romé de l'Isle, Haüy, and Dolomieu, each formulated a similar set of ideas about mineral taxonomy. Fixed and distinct species were assumed, defined primarily in terms of invariant formal relationships and constant chemical composition. With the *molécule intégrante* of Haüy's theory of crystal structure becoming the seat of the mineral species, these invariances were invested explicitly in this molecular unit as the essential attributes of all mineral "individuals."

III. MINERAL SPECIES AND QUANTIFIED CHEMISTRY

The doctrine of fixed mineral species appeared at a strikingly fortuitous time with regard to the development of chemistry. For the first decade of the nineteenth century witnessed the emergence of the quantitative framework which was to be fundamental to that science thereafter. In the first part of the decade the debate over the regularities which governed the combining proportions of chemical reagents took place. At issue was the question of whether substances could unite to form only a few compounds, each characterized by fixed weight proportions, or whether they could combine chemically, in principle at least, in a continuum of different weight proportions. The chemist Joseph Louis Proust was the principal upholder of the first view; his chief antagonist was the eminent chemist, Claude Louis Berthollet.

Their debate was "settled" by the appearance of John Dalton's atomic theory, first outlined in the third edition of the Scottish chemist Thomas Thomson's textbook, *A System of Chemistry* (1807) and then by Dalton himself in his *A New System of Chemical Philosophy* (1808). With the publication of a French

---

[36] *Cf.* above, footnote 19. A Lacroix,, *Bull. Soc. fran. Min.* **67** (1944): p. 56 ff.

[37] In fact, Haüy was given scant mention in the *Philosophie minéralogique*; Dolomieu's acknowledged predecessors were Romé de l'Isle and Werner. D. Dolomieu, *Sur la philosophie minéralogique et sur l'espèce minéralogique* (Paris, Bossange, Masson & Besson, 1801), pp. 14–17.

[38] *Ibid.,* p. 63. See also, R. Hooykaas, *Centaurus* **5** (1956–1958): pp. 317–318. Although Dolomieu did not introduce this definition of "individual" directly in answer to Daubenton, he was aware of the latter's views on mineral species, and discussed them (without naming Daubenton) in the section just prior to his discussion of the mineralogical "individual" problem. D. Dolomieu, *Sur la philosophie minéralogique,* pp 32–34. While it is reasonable to assume that Haüy and Dolomieu came to their focusing upon the *molécule intégrante* independently, Haüy did give Dolomieu credit for priority in taking the *molécule intégrante* as the exemplar of the mineral species. *Muséum d'Histoire Naturelle MS 1400* quoted in my article in *Arch. int. hist. sci.* **23** (1970): footnote 35.

[39] D. Dolomieu, *Sur la philosophie minéralogique,* p. 42.

translation of Thomson's textbook in 1809, Dalton's theory became widely known on the Continent.[1] Dalton's chemical atomism gave support to Proust's position that chemical combination could only occur in fixed weight proportions and extended this position dramatically by means of the law of multiple proportions.

With Dalton, the concept of the chemical molecule became central to chemical thought. He succeeded in quantifying molecular composition by making correspondence between the weight proportions of the constituents of a compound and the number of their atoms in the compound molecules. Dalton himself was aware of the emphasis and clarity with which he was endowing the molecular concept, as his complaint about loose employment of terms like "molecule" and "particle," cited in chapter one, bears witness. But it will also, be recalled that the one molecular definition which had impressed Dalton—John Murray's "integrant particle"—had been around for a long time when Dalton was penning his complaint and it had been from the first as much a chemical as a structural concept, even if it was most strikingly employed in the theory of crystal structure. From the time of Macquer's clear definition of the molecule given in his *Dictionnaire* onwards, the definition of the *molécule intégrante* or some similar term as the unit of composition was a commonplace of chemistry textbooks. But in an important respect, Dalton's querulousness over the molecular concept was well founded. He had complained that even the chemists who gave a satisfactory definition of "molecule" or "particle" made no use of this concept. That is, chemists before Dalton, while subscribing in general to the molecular view of material structure, that matter was made up of discrete molecular units, were not much interested in interpreting their data in terms of molecular composition.

A general reason for the opacity of pre-Daltonian chemists to the molecule as a compositional unit is inferrable from recent studies of Professor Arnold Thackray. Thackray has shown that the main thrust of chemical theory in the late eighteenth century was towards the development of a dynamical chemistry. Chemical theory was orientated towards discovering the principles which governed the forces of chemical affinity which operated between the material particles. The gravimetric information which was being thrown up in increasing amounts was to be enlisted in the program to quantify these chemical forces; ideas on the regularities governing the combining proportions of chemical reagents were substantiated by reference to the actions of chemical forces or these ideas were even conditioned by the theory of how these forces were believed to act. Chemical theory of the late eighteenth century concentrated on the forces which brought molecules into existence or tore them apart, not on the molecules themselves. This emphasis was very different from the static, molecular orientation of Haüy's and Dolomieu's doctrine of fixed mineral species—and later, of Dalton's chemical atomic theory.[2]

This point was brought out clearly in the positions developed by Proust and by Berthollet on combining proportions. While both subscribed to a molecular conception of matter, neither interpreted his data in terms of molecular composition. And with Berthollet, the dynamical orientation—figuring much more prominently in his thought than in Proust's—came into sharp confrontation with the molecular-centered doctrine of fixed mineral species over the question of fixed versus variable combining proportions.

In 1801, the same year that Dolomieu and Haüy published their mineralogical works, Berthollet produced a long paper in which he established his position that chemical combination could occur in a continuum of different weight proportions. By this time, Proust had already published a number of studies on oxides, sulfides, and other metallic compounds in which he had claimed that there existed only two oxides etc. per metal, characterized by him as the *maximum* and *minimum*, according to the proportion of oxygen to fixed weight of the metal. Each of these compounds was formed at a fixed weight proportion and there were no compounds at weight proportions intermediate to the *maximum* and *minimum* proportions.[3]

---

[1] T. Thomson, *A. System of Chemistry* (3rd ed., 5 v., Edinburgh, Printed for Bell and Bradfute, and E. Balfour, 1807) **3**: pp. 424 ff.; J. Dalton, *A New System of Chemical Philosophy* (2 v., London, R. Bickerstaff, 1808 and 1810, 1827), 1, 1: especially pp. 211–216; T. Thomson, *Système de chimie*, trans. J. Riffault (9 v., Paris, Bernard, 1809).

[2] A. W. Thackray, *Atoms and Powers* (Cambridge, Mass., Harvard University Press, 1970). There was an important interface between chemical affinity theory and crystallography stemming from a suggestion made by Buffon that chemical affinity force might be identical with universal gravitation and that molecular shape might be the variable which accounted for the selectivity of short-range forces of chemical affinity. Condorcet, for example, wrote:
"Jusqu'ici l'on n'a pu découvrir sur les lois de l'attraction à des trés-petites distances. C'est dans l'examen des phénomènes de la cristallisation que l'on pourra trouver un jour ces lois. . . . M l'abbé Haüy vient de donner, sur la formation des cristaux plusieurs mémoires que ont répandu un grand jour sur cette matière importante. Cependant on est peut-être encore bien éloigné d'en savoir assez pour pouvoir y appliquer le calcul, et connaître les lois de la force attractive que préside à la cristallisation." M.J. A.N.C., Marquis de Condorcet, Note "Haüy," "Notes sur Voltaire," *Œuvres de Condorcet*, ed. A. Condorcet O'Conner & M. F. Arago (12 v., Paris, Firmin Didot, 1847–1849) **4**: p. 425. I thank Prof. Henry Guerlac for calling this to my attention. Cf. also my article, "Thomson Before Dalton," *Annals of Science* **25** (1969): pp. 233–236 for Thomas Thomson in this same context.

[3] C. L. Berthollet, "Recherches sur les lois de l'affinité," *Paris, Mém. de l'Inst.* **3** (1801): pp. 1–96; J. L. Proust, "Recherches sur le bleu de Prusse," *Jour. de phys.* **45** (1794): pp. 334–341. S. Mauskopf, "Proust, Joseph Louis," *Dictionary of Scientific Biography* **11**: pp. 166–172.

To a degree, both Proust's and Berthollet's positions had their origin in eighteenth-century theory of chemical affinity. Each chemist elaborated upon some aspects of this general dynamical background and modified others in order to form and justify his position. Proust had formulated his belief in definite proportions first and, in a sense, his was the more traditional stance. The main idea of the theory of chemical affinity had been that the reactivities of chemical substances were the result of attractive forces operating between the molecules of these substances much as gravity exerts its attraction between massive bodies. The "intensity" of the force between any two substances was assumed to be constant and unaffected in its action by any physical or chemical factors. At least the relative strengths of affinity which different reagents exercised towards a given substance and which could be determined by the order in which these reagents replaced one another in combination with the substance were considered to be invariant before Berthollet.

A number of eighteenth-century chemists had drawn gravimetric implications from chemical affinity theory, espousing a doctrine of definite combining proportions which might be called the "principle of constant saturation proportions." According to this principle, two reagents, whether they be an acid neutralizing a base, or water dissolving a salt, were thought to have one and only one weight proportion at which true chemical combination took place: that at which their forces of chemical affinity were "saturated." That is, the chemical affinity force of a given weight of one reagent was just capable of attracting a certain weight of a second reagent into chemical combination. At this point, there would be "saturation" of the affinity force. Since the strength of this force was considered to be invariant, the weight proportions at which saturation took place had to also be constant.[4]

There were serious anomalies in this principle. The amount of solute a solvent could dissolve varied with temperature; acidic and basic salts could be formed as well as neutral salts in a neutralization reaction; and, especially after the pneumatic chemistry of Lavoisier clarified the elementary make-up of compound substances, it was apparent that there were many elements which could form stable compounds at more than one combining proportion. To the first of these objections, it was claimed that the amount of caloric or "matter of heat," in the solution as measured by its temperature, entered into chemical combination with the solute and solvent and consequently changed the nature of their affinity force. The other difficulties were circumvented by the suggestion that there was still only one true saturation proportion between a given set of reagents, and the additional stable proportions were in fact new combinations of the compound at saturation proportions with some excess of one of the reagents.

It is impossible to determine whether Proust was aware of the principle of constant saturation proportions since he made no reference to any of its formulations. Nor was he in general inclined to theorize much about why chemical combination occurred in fixed proportions. The justification he gave for his own belief in this was most often a combination of experimental data with an appeal to a "balance of nature" which governed and insured proportionality in chemical combination. Moreover, Proust's position on definite proportions differed from that of the proponents of constant saturation proportions in one significant way: Proust took for granted that there existed multiple stable combining proportions between reagents, rather than positing one "true" saturation proportion and trying to explain the other proportions in terms of that one.

Yet chemical affinity was as much Proust's conceptual framework as it had been his predecessors; when he did refer to chemical theory to support his belief in fixed combining proportions, it was to the saturation of affinity forces that he automatically—if vaguely—appealed. For example, in response to Berthollet's contention that combination could take place in a continuum of weight proportions, Proust wrote:

. . . a molecule of potash, of an earth, or an oxide which is found in the presence of an acid attracts neither half nor a quarter of what it needs for saturation. At the moment of contact, it forms a complete combination, obedient to the proportions which its affinities assign to it. This is the way of chemical combination in general.[5]

Without doubt there was a broad similarity between Proust's belief in definite combining proportions and the doctrine of fixed mineral species. This lay in the mutual assumption of fixity and discreteness of type among inorganic substances and the common definition of this type in terms of fixed chemical composition. Yet there were also deep differences, having to do with divergencies in their scientific interests and with a fundamental difference in their theoretical orientation.

As one would expect from a chemist, most of Proust's work on combining proportions dealt with relatively simple laboratory substances such as metallic oxides

---

[4] S. Mauskopf, *Annals of Science* 25 (1969): pp. 229–242. In the immediate post-Daltonian period, William Higgins neatly summarized the relation of the eighteenth-century principle to its sequel: "In short, there is a limit to the proportions in which the particles of elementary matter, as well as those of atoms, unite, which the old chemists expressed by the term saturation, and the modern ones by that of definite proportions." *Experiments and Observations on the Atomic Theory* (Dublin, Graisberry & Campbell, 1814), p. 14.

[5] J. L. Proust, "Sur les oxides métalliques," *Jour. de Phys.* 59 (1804): p. 329. Cf. also, "Recherches sur le cuivre," *Annal. de chimie* 32 (1800): p. 32, "Mémoire pour servir à l'histoire d'antimoine," *Jour. de phys.* 52 (1802): p. 332.

and sulfides. Regarding them, Proust tried to show that each pair of constituents could form two or at most only a few distinct binary compounds. Being mineralogists, on the other hand, Haüy and Dolomieu were interested in classifying more complex mineral substances, concentrating on crystal form as the decisive criterion for such classification. Therefore Haüy and Dolomieu said little about the question of how many possible combining proportions could exist between two reagents—a chemical question—and Proust in his turn was unconcerned with distinguishing the essential form-giving component of a mineral from the accidental ones—a question of mineral taxonomy.

This difference of interest between Proust and the mineralogists was reflected at the theoretical level. Proust, with his background in chemical affinities, had no conception of anything like the static, geometrical *molécule intégrante* which was the locus of the mineralogists' species.[6] Nor was there any obvious way in which this concept could have been particularly useful to him, given the different kinds of questions with which he was grappling. The theoretical situation on the part of Haüy and Dolomieu was somewhat more complicated, for both of them recognized that the formation of the *molécule intégrante* itself was due to the action of chemical affinity forces. But questions of how the essential compound of their mineral came into existence or what other compounds its constituents could form were completely subordinate to the issue of determining that this compound was the essential one in the mineral—that the formal and compositional invariances of the mineral sample reflected the form and composition of the essential *molécule intégrante*.

The difference—and the relation as well—between the doctrines of fixed species and definite proportions were strikingly displayed in the circumstances which brought about Proust's most emphatic statement of his law of definite proportions and his most elaborate justification of it. This defense of definite proportions was contained in a paper Proust published in 1806, at just the period when his famous polemic with Berthollet over combining proportions had reached its height. Most historians have quite naturally assumed that this particular statement and defense of the law of definite proportions were part of that polemic.

Curiously enough, however, the paper was elicited not by any criticisms of Berthollet or anyone else of Proust's views on combining proportions, but by a critique of Haüy's and Dolomieu's doctrine of fixed species. The critic, Jean François D'Aubuisson, was a disciple of the German geologist and mineralogist, Abraham Gottlieb Werner. D'Aubuisson had published a series of attacks on Haüy's and Dolomieu's species concept beginning in 1802, in which he had argued against the existence of the sharp divisions and essential characteristics in minerals which their mineralogy posited; the best one could hope to do rather was what Werner had already done in supplying a descriptive and practical system of mineral classification based on the external characteristics. It was futile to aim at the exactitude of fixed species. As justification, D'Aubuisson appealed to chemistry, arguing in a manner similar to Berthollet's that chemical combination did not necessarily take place in fixed proportions.[7]

It was this which called forth a reply from Proust. In attacking the doctrine of fixed species, D'Aubuisson had cited some mineral analyses of the chemist, Martin Heinrich Klaproth:

The analysis of *gray copper* which Klaproth has recently given are a new example of combinations in variable proportions.[8]

To which Proust retorted:

Gray copper, I answer, does not belong at all to that order of *combinations* which chemists are examining at this time for unravelling the principles of their formation. A combination, according to our principles, Klaproth would say to you, is silver sulfide.... this is a privileged product to which Nature assigns fixed proportions.[9]

Since D'Aubuisson was talking about minerals, Proust —unusually for him—devoted a good deal of attention to expounding his ideas on mineral chemistry. Here the difference between Proust and Haüy was most striking. For Proust's position was that minerals were made up of mixtures of his simple binary compounds, such as metallic oxides and sulfides and the like, rather than of one essential compound with accidental intermixtures of other substances. It was at the level of the binary compound that definite proportions held in the inorganic kingdom.[10]

The overlap between the fixed species doctrine and definite proportions was obviously close enough for Proust to have assumed that D'Aubuisson *must* have been attacking his views. But the strange feature of Proust's defense was its total obliviousness to the species doctrine which D'Aubuisson was really attacking. There is not the slightest hint that Proust consciously

---

[6] Interestingly enough, Haüy himself set aside his *molécule intégrante* in favor of a dynamical model when considering chemical (as opposed to mineralogical or crystallographical) problems. *Cf.* my article, "Haüy's Model of Chemical Equivalence; Daltonian Doubts Exhumed," *Ambix* 17 (1970): pp. 182-191.

[7] J. F. d'Aubuisson, "Lettre de J. F. D'Aubuisson à J. C. Delamétherie sur quelques points de minéralogie," *Jour. de phys.* 54 (1802): pp. 333-344, 414-420; "Tableau de la classification des minéraux"; *ibid.* 60 (1804): pp. 171-178, 329-333; "Lettre de M. d'Aubuisson à M*** sur quelques objets de minéralogie," *Annal. de chimie* 57 (1806): pp. 273-316.

[8] *Ibid.*, p. 296, fn. 1.

[9] J. L. Proust, "Sur les mines de cobalt, nickel et autres," *Jour. de phys.* 63 (1806): p. 367.

[10] *Ibid.*

was aware of the similarity between his views and those of Haüy and Dolomieu or concerned about their mineralogy. Indeed, in what might be the one oblique reference to the species doctrine in this paper, Proust commended D'Aubuisson for having pointed out "the futility of certain definitions."[11]

Proust's chief adversary over the issue of combining proportions, Berthollet, was not so oblivious to the doctrine of fixed mineral species. Berthollet was a much more powerful theorist than Proust; his ideas on combining proportions were embedded in a general framework of chemical dynamics and he was quick to read into the *molécule intégrante* and fixed species a different approach from his own for explaining chemical reactivity.

In line with his general program for working out a theory of chemical dynamics, Berthollet had been led to investigate exceptions to that general assumption of the eighteenth-century theory of affinity that the operation of chemical affinity forces was invariant regardless of the circumstances in which the chemical reaction took place. These anomalies had to do with the corollary of this assumption, that the direction of a replacement reaction was determined solely by the order of the affinity strengths of the competing substances.

Berthollet came to see the operation of chemical affinity forces as being much more complex than the traditional theory of affinity had supposed. In the case of the direction of a replacement reaction, for example, such factors as the masses of the reacting substances and their physical states before and after the reaction appeared to have an effect on the way the chemical reaction would proceed as well as did the affinity strengths. In developing these ideas, Berthollet was led to reject another implicit assumption of affinity theory: that the replacing substance expelled all of its competitors in a replacement reaction. In place of this Berthollet put forth the notion of a dynamical equilibrium. A replacement reaction never went totally to completion—one substance never completely expelled another from its combination with a third substance. Rather, the first two shared the third substance in a proportion governed by the amounts of the two competitors present at the scene of the reaction and their relative strengths of affinity for the third substance.[12]

From these dynamical considerations Berthollet went on to draw a conclusion about the possible combining proportions between reactants. Two competing reactants could combine with a third substance in a continuum of different weight proportions. He generalized this, moreover, beyond the results of replacement reactions into a principle of combining proportions valid for all chemical reactions.[13]

Berthollet's belief in a general principle of variable proportions was fostered not only by his modification of elective affinity mechanism, but even more by his belief that solutions, such as salt in water, were true chemical combinations, albeit weak ones. This idea in itself was not novel with Berthollet; solutions had been considered to be some sort of chemical combination—the result of forces of chemical affinity—by virtually all eighteenth-century chemists and had even been accommodated to the principle of constant saturation proportions. What Berthollet did was to wrench this class of chemical substances out of this accommodation, accept the variability of the possible proportions of solute to solvent, and appeal to solutions as a paradigm of chemical combination:

> One consequence of the preceding observation is, that we must follow the same laws in combination as have been observed in the chemical action which produces dissolution.[14]

Just as water could take up a continuum of proportions of salt until saturation, and, at each proportion, give an homogeneous solution, so other chemical substances presumably could unite in a continuum of proportions up to that of saturation.

While Berthollet recognized that many combinations seemed to occur only in certain fixed proportions, he explained these as being due to certain physical attributes of the compounds at these proportions—such as greatest condensation, or volatility—which rendered them more easily removable from the scene of chemical activity by precipitation, crystallization, or volatilization. Fixed proportions, then were to Berthollet in no sense a defining mark of chemical union, but rather was the result of physical attributes of the compound at those particular proportions, which facilitated the removal of it from further influence of the reactants.[15]

It was this position, sketched out by Berthollet in the *Récherches* of 1801 and developed in the *Essai* of 1803, that led to a famous polemic with Proust over combining proportions. But, from this same standpoint, Berthollet also attacked Haüy's and Dolomieu's doctrine of fixed mineral species. Indeed, Berthollet's confrontation with this doctrine took place *before* his controversy with Proust over the law of definite proportions. For, in the *Essai* itself, Berthollet used his

---

[11] *Ibid.*

[12] Berthollet's dynamical chemistry was worked out most elaborately in his *Essai de statique chimique* (2 v., Paris, Firmin Didot, 1803). *Cf.* S. C. Kapoor, "Berthollet, Proust and Proportions," *Chymia* 10 (1965): pp. 53–110; A. W. Thackray, *Atoms and Powers*, pp. 199–233.

[13] First in *Paris, Mém. de l'Inst.* 3 (1801): pp. 13–14 and then in the *Essai de statique chimique* 1: pp. 20, 62 ff,

[14] *Ibid.* 1: p. 60. Translation from *An Essay of Chemical Statics*, trans. B. Lambert (2 v., London, J. Mawman, 1804) 1: p. 34.

[15] C. L. Berthollet, *Essai de statique chimique* 1: pp. 63–67, 337 ff., 373–380 (summary).

arguments in favor of variable combining proportions—not against Proust, though he did criticize Proust's assertions about the limited number of oxides which metals could form [16] but rather against Haüy and Dolomieu. In a chapter significantly entitled, "Des proportions des élémens dans les combinaisons," he launched his attack against them:

> In fine, latterly, it has been found that the form of the moleculae of a substance, or of the integrant parts of a combination, determined all the secondary forms which could be produced by their union, and it has been inferred that this primitive form determined the combinations themselves, and consequently the proportions of their elements.[17]

As this quotation indicates, Berthollet's attack on Haüy and Dolomieu was pitched at a theoretical level, involving the nature and function of the *molécule intégrante*. In Berthollet's interpretation of their ideas, chemical combination was somehow determined by the shapes of the constituent particles, and the geometrical constancy of molecular shape (revealed in the fixity of crystalline form) conditioned the constancy of the resultant compound's chemical composition.[18]

All of this, in fact, went far beyond anything Haüy (or Dolomieu) had explicitly asserted. Haüy's emphasis had been on the already-existent compound molecule, with little concern with its manner of formation. But Berthollet was interested in the mechanics of chemical combination in the *Essai;* hence he dealt with Haüy's doctrine of fixed mineral species as if it were a theory about chemical reactions and the formation of chemical compounds. He tried to refute the contention, which he read into Haüy's doctrine, that chemical combination was a kind of geometrical additive process.

Berthollet's argument against Haüy here brought into relief his dynamical orientation to chemical theory. While he did subscribe to a molecular theory as a matter of course, nevertheless, his chemical theory was less concerned with the molecules as such than with the interplay of intermolecular forces of chemical affinity. His molecules were functionally little else than bundles of affinities; chemical phenomena were fundamentally ever-changing patterns of affinity forces.

With force as the focus for Berthollet, other molecular attributes, such as shape, were secondary and conditioned by it. Molecular shape was an emergent quality to Berthollet, which came into being in the process of combination (and condensation) resulting from the interplay of caloric and affinity forces. Hence, he reversed the order of precedent which he read into Haüy: molecular shape, rather than determining chemical combination, was itself determined by it. In support, he denied that molecules even had any determinate shape when they were widely separated from each other—in gases or liquids.[19]

If shape was dethroned as the primary molecular attribute, the privileged state which Haüy had accorded to crystal form in his mineralogy was obviously fair game. What applied to molecular shape must hold even more for the arrangement of molecules into crystals. And, in fact, to Berthollet, crystallization was the result of the weakest of all intermolecular forces, only taking place under the gentlest of conditions. Berthollet was not arguing against the notion that crystals were composed of the orderly arrangement of determinately shaped molecules. Rather, what he was asserting was that both the molecular shape, and the arrangement of molecules to form crystals—crystal shape—were brought about by chemical affinity forces, and in no way determined the nature of the chemical combination.[20] A compound which crystallized poorly, or not at all, was, after all, still a compound and retained all its chemical and physical properties.

In a long note appended to the main body of the chapter on chemical proportions in the *Essai,* Berthollet pressed his attack on Haüy's species doctrine.[21] Essentially, his critique centered upon two issues: the criteria used to classify minerals into species, and the very existence of distinct and separate species. With regard to the first issue, Berthollet down-graded the primacy Haüy had given to crystallography, in favor of chemistry. Here, Berthollet was bringing to the fore an issue which was to dominate discussion in mineralogy for the next half-century: which of the developing physical sciences afforded the most accurate classification scheme for minerals? In support of chemistry (against crystallography), Berthollet was able to cite a significant number of anomalous variations of the correlation which Haüy had drawn between form and composition. These were to plague Haüy for the next twenty years. There were the known cases where the same chemical substance crystallized in different geometrical and incompatible crystalline forms—what was subsequently to be termed polymorphism—and there were the cases where substances of different or variable composition assumed the same crystalline form—to be known as isomorphism.

As to the second issue, the existence of fixed mineral species characterized by definite proportions, Berthollet's position had to be, and was, negative. In considering this issue, especially from the chemical point of view, Berthollet brought Haüy's mineralogy

---

[16] *Ibid.* **2**: pp. 369–382.
[17] *Ibid.* **1**: p. 336. Whole chapter is 334–386. Lambert trans., **1**: p. 249.
[18] *Ibid.,* (*Essai*), **1**: p. 384.
[19] *Ibid.* **1**: pp. 380–381.
[20] *Ibid.* **1**: p. 383 for argument that crystallization took place only under gentle conditions; p. 439 for more general relation of crystal form to chemical combination.
[21] *Ibid.* **1**: pp. 433–449 (note XIV). Dolomieu was included here with Haüy.

close to Proust's chemistry by answering the one in ways similar to those by which he was subsequently to reply to the other. Berthollet realized that Haüy's way of establishing his doctrine of fixed mineral species was to assume a criterion for chemical "purity" in crystal form. Berthollet answered Haüy on this point with one of his rare attempts to define some criteria of his own for true chemical composition: by which those substances which did not manifest fixed proportions—glasses and presumably complex isomorphic minerals in the case at hand—could still be included under this rubric:

> The uniformity in the composition, notwithstanding the difference of the specific gravity of the elementary parts; the transparency, which proves that they no longer exercise a separate action on the rays of light; the properties which are in common, but different from those of the separate elementary parts, are nevertheless undeniable indications of combination.
>
> All the characters of a combination are unquestionably found in glass, which may be composed of very different proportions, and it cannot be asserted that this combination has determinate proportions, and a form which belongs to its integrant particles, and that all the rest is interposed without entering into the formation of the compound. What I now observe of glass is applicable to all the transparent minerals which contain oxides or other elements foreign to those to which the form of their integrant moleculae is attributed.[22]

Thus, after upholding chemistry as the determinative science in mineral taxonomy, Berthollet then proceeded to deny it the power to define mineral species, for the very good reason that such species, defined by fixed chemical composition, could not exist for him.[23]

The five years following the appearance of Berthollet's *Essai de statique chimique* witnessed the shift in focus of Berthollet's attack from the mineralogists to Proust, the chemist, in the more famous polemic between them over fixed versus variable proportions. During the same period Haüy's mineralogy found sympathetic defenders: the Irish chemist Richard Chenevix against the Wernerians, and the mineralogist Alexandre Brongniart against Berthollet.[24] Haüy himself gave a detailed reply to Berthollet in a work whose purpose was to appraise the success of his correlation between crystal form and chemical composition. This was his *Tableau comparatif des résultats de la cristallographie et de l'analyse chimique,* published in 1809. The *Tableau* was in fact more than an impassive appraisal; it was a defense of his fixed species concept and an assertion that in matters of disparity crystallography was to take precedence over chemistry.

Haüy was forced to the defense not only by Berthollet, but by other critics of his mineralogy, and even criticism of his theory of crystal structure. In particular, the cases incompatible with his correlation of form and composition which had already been cited by Berthollet in his *Essai* continued to attract attention. The central issue of the *Tableau* was Berthollet's critique of the species concept. Haüy blithely argued through rather than at Berthollet; that is, he ignored (or overlooked) Berthollet's contention that there could not and did not exist fixed mineral species, and he turned round the disparities between form and composition which Berthollet had hurled at his mineral classification, by asserting that such disparities pointed up, more than ever, the primacy of crystallography in the determination of mineral species.[25]

However, Haüy was not here accepting the existence of such disparities; on the contrary, he clung more tenaciously than ever to the strict correlation he had drawn between crystal form and chemical composition. But, in defending this correlation, Haüy himself was constrained to make explicit the geometrical chemistry which Berthollet had read into his species doctrine in the *Essai de statique chimique*:

> I avow, however, that it seems most likely to me that the integrant molecule is one in its composition as well as in its form. It would be difficult to conceive how the first could vary among individuals of the same species while the second undergoes no change, if one supposes, as is most natural, that the elementary molecules themselves have determinate shapes.[26]

While even here he never actually said that chemical combination involved simply a geometrical juxtaposition of elementary molecules to form the compound one, this is clearly the direction of thought towards which this statement was pointing. Haüy himself recognized its importance in later years; he styled it his version of the law of definite proportions and indeed the parallel and precursor of Dalton's atomic theory.[27]

Precursor it was not, for by 1809, John Dalton's atomic theory had already made its public appearance and had received experimental support. During the whole period of the debate over fixed versus variable combining proportions, Dalton had been slowly—hesitantly at first and then more surely—extending his newly created concept of relative atomic weight, in-

---

[22] *Ibid.* 1: p. 439. Lambert trans., 1: pp. 438–439. Berthollet used the same argument later against Daltonian stoichiometry in his *Introduction* to the Riffault translation of Thomson's *System of Chemistry* (above, footnote 1 for bibliographical details), 1: p. 24.

[23] C. L. Berthollet, *Essai de statique chimique* 1: pp. 445–446.

[24] R. Chenevix, "Reflections on Some Mineralogical Systems," *The Philosophical Magazine* 36 (1810): pp. 286–303, 378–391, 413–423; 37 (1811): pp. 39–51, 117–128. He was answered by Thomas Thomson for the Wernerian position. "Some Observations in Answer to Mr. Chenevix's Attack on Werner's Mineralogical Method," *Ann. Phil.* 1 (1813): pp. 241–258. A. Brogniart, *Traité élémentaire de minéralogie* (2 v., Paris, Deterville, 1807) 1: pp. 46 ff.

[25] R. J. Haüy, *Tableau comparatif,* p. x.

[26] *Ibid.,* pp. x–xi.

[27] R. J. Haüy, *Traité de cristallographie* (3 v., Paris, Bachelier et Huzard, 1882) 2: pp. 413–414.

vented to explain physical characteristics of air and of gaseous solutions, to questions of chemical composition.[28]

With Dalton, the potentialities of the chemical molecule for interpreting and ordering chemical data began to be realized. But—Haüy notwithstanding—it was weight and not form which was the instrumental parameter. Dalton's achievement was to make explicit and develop the idea that matter was made up of a fairly large but finite number of distinct elementary substances, the atoms of each of which had fixed weights characteristic of the element. Chemical combination consisted in the union of a definite number of atoms of each reagent to form a "compound atom" or molecule. Invariance of the atomic makeup of the molecule insured the invariance of the gravimetric composition of the compounds.[29] The question of why chemical combination occurred was shunted aside by Dalton in favor of the question of "how many atoms," though the search for the cause of chemical reactivity was never very far from most chemists' minds and was soon reintroduced under new guises.

Dalton did not presume to ascertain the absolute weights of atoms, but rather their relative atomic weights and these from the available gravimetric data. The means to do this involved the use of certain simplifying assumptions about the numbers of atoms in a molecule. The fundamental one was that, if only one compound between two elements was known to exist, one was to assume that its molecule was a binary one, made up of one atom of each element. Water was such a binary compound for Dalton. In this case, the relative atomic weights of hydrogen and oxygen could be inferred directly from their gravimetric proportions in water, which was about one to seven. If more than one compound was formed, it was assumed that more complex ternary, quaternary, etc. molecules were formed out of two atoms of one element to one of the other, two to two, etc.[30]

It was these additional assumptions about the molecular make-up of compounds which enabled Dalton and his followers to give pattern to much of the accumulated gravimetric data. This they did by what became known as the law of multiple proportions. According to this law, the weight ratios of one reagent forming several compounds with another are as small integers when the weight of the second reagent is made constant. This law, so awkward to express verbally, was extremely easy to visualize in terms of Dalton's molecular composition, where these integral weight ratios of the first reagent expressed the numbers of its atoms present in the molecules of the different compounds. Dalton applied this law to compounds of nitrogen, carbon, and sulfur with oxygen and to a few hydrocarbons in the first part of *A New System of Chemical Philosophy*. It was extended dramatically to various salts by Thomson, Wollaston, and, most notably, Jöns Jacob Berzelius.[31]

Haüy was certainly correct to claim priority over Dalton in emphasizing the molecule as the locus of chemical composition at constant proportions. But the differences in their conceptual contexts made their molecular concepts far from parallel. For Haüy, the *molécule intégrante* was a fully formed unit of structure. Structural and compositional invariances became entwined in his mineral species concept, but Haüy was never very concerned with working out the details of how these invariances might be related. It sufficed for him to assume that the elementary molecules which composed his *molécules intégrantes* were determinate and invariant in shape and that their shapes somehow determined the form of the compound molecule.

Dalton, not being a crystallographer, did not share Haüy's interest in the molecule as a structural unit; for him it was a unit of composition. His very image of the atoms—globular in form because of their attendant caloric atmosphere—and of the molecule as an agglomeration of such atoms was different from Haüy's molecular polyhedra.[32] Potentially, however, Dalton's atoms might be complementary to Haüy's *molécule intégrante*. If Daltonian stoichiometry could provide information about the number of atoms in the compound molecule, Haüy's structural theory could give insight into its form. But, despite their obvious knowledge of each other's achievement, neither Haüy nor Dalton was the one to bring their theories into fusion. It was rather their younger contemporaries and their successors who were to attempt this. Within a decade of the appearance of Dalton's atomic theory, the two main areas of interface between chemical atomism and molecular crystallography were delineated. One was the

---

[28] A. Thackray, *John Dalton* (Cambridge, Mass., Harvard University Press, 1972), chaps. 5 and 6.

[29] Cf. J. Dalton, *A New System of Chemical Philosophy* 1, 1: pp. 142–143 where Dalton made just this point in arguing against Berthollet's doctrine of variable proportions. Dalton was also arguing here that the atoms themselves of each element were invariant.

[30] *Ibid.*, pp. 213 ff. Dalton possibly had reached this calculating procedure for water and for the oxides of nitrogen by the autumn of 1803. A. Thackray, *op. cit.*, pp. 75 ff. and 100.

[31] J. Dalton, *A New System of Chemical Philosophy* 1, 1: p. 219. Dalton had, however, by 1807 already extended his law of multiple proportions to a much wider group of chemical combinations. C. A. Thackray, *op. cit.*, pp. 111–114. T. Thomson, "On Oxalic Acid," *Phil. Trans.* 98 (1808): pp. 63–95; W. H. Wollaston, "On Super-acid and Sub-Acid Salts," *ibid.*, pp. 96–102; For a convenient summary of Berzelius publications on combining proportions, commencing in 1810, *cf.* J. R. Partington, *A History of Chemistry* (4 v., London, Macmillan 1962–1970) 4: pp. 153 ff.

[32] J. Dalton, *A New System of Chemical Philosophy*, 1, 1: p. 147. But see below, chap. 4 for Dalton's comments on the agglomeration of atoms to form crystals.

question of mineral taxonomy, particularly the mineral species question. The other was the even broader question of determining the arrangements of atoms in their molecule for the purpose of providing a structural theory of chemical reactivity.

The question of mineral species was the one closer to Haüy's own interest, as the *Tableau comparatif* bore witness. The main issue there had been whether fixed mineral species existed and which science—crystallography or chemistry—was to be the ultimate arbiter of mineral taxonomy. These issues continued to dominate the mineralogical discussions but, from the chemists' point of view at least, took on new interest with the impact of Daltonian atomism.

In 1814 Berzelius published a sketch for a mineralogical system which he dedicated to Haüy.[33] By this time, he had already made his mark as one of the most important younger chemists by his work in electrochemistry and in quantitative analysis. In this latter field, he had corroborated Dalton's laws by massive systematic analysis of inorganic compounds. It was to benefit mineralogy by these recent advances in chemistry—so much his own doing—that Berzelius proposed his new system.

Indeed, he meant to go further: he wished to make mineralogy "merely a part or an appendix to chemistry."[34] For the higher taxonomic divisions, he brought to bear his dualistic electrochemistry; for the lowest, the species, Daltonian atomism. The result, for the species concept, was that, although Berzelius asserted the primacy of chemistry in mineralogy as vigorously as had Berthollet, he arrived at the same doctrine of fixed mineral species as had Haüy. Out of the atomic theory now appeared the same principle of fixity of species which the French crystallographers had for so long asserted. Thus, Berzelius drew a distinction between organic classification and that of minerals: for the former, transitions between species might well be possible, but for the latter, they were impossible and "the cause of this phenomenon is found in the very principle which presides over combinations in the inorganic realm"—Dalton's law of multiple proportions.[35] In his actual definition of mineral species, Berzelius consciously echoed Haüy:

For minerals of the same species, I understand here the same thing as Haüy, *the same composition with the same primitive form.*[36]

---

[33] Originally, published in Swedish in 1814. I have used the French edition: *Nouveau système de minéralogie,* trans. n.n. (Paris, Mequignon-Marvis, 1819).

[34] *Ibid.,* p. 4.

[35] *Ibid.,* p. 118. Berzelius's sharp distinction between the organic and the inorganic realms was accentuated by his belief in a vital force. *Ibid.,* p. 117. Transitions were possible in the organic kingdom but not among minerals. Berzelius defined transitions, as had Haüy, as mechanical mixtures of one compound in another. *Ibid.,* p. 120.

[36] *Ibid.,* p. 44.

But, despite his agreement with Haüy, Berzelius's chemical orientation resulted in differences in emphasis, the use of "primitive form" rather than "integrant molecule" being an indication. Although Berzelius agreed that correlation of crystal form was a good criterion for species determination, he refused to accord to crystallography the dominance with which Haüy had invested it, if this be to the detriment of chemistry.[37] And this issue was shortly to become an acute one in mineralogy.

The whole question of the priority of one science or another in species determination arose for a reason we have already touched upon, but which now requires a closer look. This was the fact that the strict correlation which Haüy had drawn between his two species criteria, crystal form and chemical composition, did not hold generally. Haüy had been aware of exceptions to it from the start of his concern with mineralogy; indeed, it was precisely the appearance of anomalies that led Haüy, in the first edition of the *Traité de minéralogie,* to assert the primacy of crystallography over chemistry, in the first place. For Haüy blamed the discrepancies to his correlation on the inadequate state of chemical analysis. This included not only the chemists' inability to give reliably precise analyses—a just point for the 1790's and early 1800's, and one that plagued the controversy between Berthollet and Proust. But more theoretically, Haüy claimed that chemistry, in its current state, was unable to distinguish in a mineral those substances which were indeed chemically combined, and which constituted the chemical essence of the mineral, from those which were incidentally mechanically intermixed. On the other hand, the constancy (or variability) of crystal form for mineral samples of roughly the same chemical composition seemed to Haüy to afford a precise way of telling whether such minerals were of exactly the same composition and hence whether at least some of the samples had chemical substances only mechanically intermixed.

In the later 1790's, at the time when Haüy was just developing these ideas, he achieved, indirectly, a striking success. The two minerals, emerald and beryl, possessed similar crystalline form and other physical attributes, except for the green coloring of emeralds. Haüy, suspecting that they were essentially identical chemically, urged his chemical collaborator, Louis Nicolas Vauquelin, to analyze both minerals to ascertain whether this was true. In the course of his studies, Vauquelin discovered a new metallic earth in beryl later named beryllia by Klaproth. Following Haüy's hypothesis about the essential chemical identity between beryl and emerald, Vauquelin re-analyzed emerald, and he discovered the new earth there as well. Thus, Haüy and crystallography had led chemistry to the

---

[37] *Ibid.,* p. 189.

discovery of a new element and to the demonstration of its presence in a substance whose analysis had heretofore seemed complete.[38]

But in the first two decades of the nineteenth century, even though techniques of chemical analysis were refined, discrepancies between crystal form and chemical composition persisted, eventually to overthrow the rigid relation Haüy had assumed between them.

There were discrepancies where constant crystalline form was associated with variable composition and others where apparently identical chemical substances crystallized in geometrically incompatible forms and exhibited different physical properties. The first type—subsequently known as isomorphism—seemed to violate the chemical criterion of Haüy's concept of fixed species, and the law of definite proportions as well. There were three main subtypes of this discrepancy. First, there were those families of complex minerals composed of silica and various alkaline earths, such as the alums, feldspars, the pyroxenes, etc. In each of these families, crystalline forms of different samples were nearly identical, but the weight proportions of the constituents, and some of the properties, varied greatly. Moreover, unlike the beryl-emerald case, Haüy was unable to discover a means of definitely separating the "essential" from the incidental constituents.

Second, there were less complex minerals which, though of different but related chemical composition, nevertheless crystallized in virtually identical forms. So long as the form in question was one of the "limiting forms"—the cube, regular octahedron, or dodecahedron—Haüy was undisturbed. But in the case of the three minerals, calcspar, bitter spar, and iron spar, the crystals of each appeared to be rhombohedra whose interfacial angles were all nearly 105°, though William Hyde Wollaston, using an improved goniometer of his invention, was able to show in 1812 that their interfacial angles did indeed differ by close to two degrees.[39] Moreover, though each of these minerals was composed primarily of a particular carbonate salt (calcium carbonate for calcspar, magnesium carbonate for bitter spar, and iron carbonate for iron spar), nevertheless all three carbonates could be found in varying proportions, in samples of all three spars. Haüy had originally thought that iron spar was really calcspar with a small addition of intermixed iron, but he was compelled to abandon this idea on the basis of chemical analysis. In the *Tableau comparatif,* he adopted and maintained thereafter an explanation originally suggested by Romé de l'Isle, that iron spar was a pseudomorph of calcspar.[40]

Finally, and most striking of all, was a phenomenon of the laboratory: the mixed crystallizations of sulfate salts of zinc, copper, and iron. Early observations of these had been made by Nicolas Leblanc in connection with his studies of the conditions under which crystallization took place. Leblanc noted that sulfate salts seemed to be able to crystallize out homogeneously together in all weight proportions; those of iron and copper sulfate, for example, always did so in the form of rhombohedral crystals—the crystalline form of iron sulfate. Leblanc considered these mixed crystallizations to be something more than mere interposition of the sulfates.[41]

These mixed crystallizations, noted by Leblanc in the monograph of his investigations into the conditions of crystallization he published in 1802, entitled *De la cristallotechnie,* did not go unnoticed. Indeed, further investigations revealed something extraordinary: very small relative proportions of iron sulfate "dominated" the crystallization of iron-copper sulfate—that is, the iron sulfate impressed its crystalline form upon the mixed sulfate crystals. The case of copper sulfate dominance in copper-zinc mixed crystallization was parallel to that of iron sulfate, and even more striking.[42] A systematic investigation of these mixed crystallizations was undertaken by a pupil of Haüy, François Sulpice Beudant, in 1816. The importance of this laboratory phenomenon to Haüy's mineral species concept is revealed in the title of Beudant's first paper on the subject: "Recherches tendantes à déterminer l'importance relative des formes cristallines et de la composition chimique dans la détermination des espèces minérales" and, in Beudant's justification for his studies:

It seems to me that one means, not of solving the problem [of what constitutes mechanical intermixture] but of throwing some light on it, would be to ascertain ... what were the limits of mixture with conservation of form; that is, to find at what point a definite compound can become intermixed with foreign substances, in which the system of crystallization which is proper to it, changes. This is the object of the experiments which I am going to describe.[43]

Beudant found that as little as 2–3 per cent of a triple sulfate crystallization need be iron sulfate for the crystals to be formed in its shape. Clearly, to call (in this

---

[38] *Cf.* R. J. Haüy, *Traité de minéralogie* (1822) **2**: pp., 522–525 for Haüy's account of this episode.

[39] W. H. Wollaston, "On the Primitive Crystals of Carbonate of Lime, Bitterspar and Iron Spar," *Phil. Trans.* part I (1812): pp. 159–162. Haüy never accepted Wollaston's measurements. *Cf. Traité de cristallographie* **2**: p., 451.

[40] R. J. Haüy, *Tableau comparatif,* pp. 279–281. In pseudomorphs, the chemical composition of the crystal has slowly changed although the original crystal form remains.

[41] N. Leblanc, *De la cristallotechnie ou essai sur les phénomènes de la cristallisation* (Paris, n.p., 1802), pp. 49–50.

[42] J. J. Bernhardi, *Gehlen. Jour.* **8** (1809): pp. 385–388. Bernhardi was opposed to Haüy's molecular theory of crystal structure.

[43] F. S. Beudant, "Recherches tendantes à déterminer l'importance relative des formes cristallines et de la composition chimique dans la détermination des espèces minérales," *Annal. des mines* **2** (1817): p. 8.

instance) iron sulfate the "essential" compound, and the other sulfates merely incidental, was verging on the absurd. Beudant himself was uncertain whether to characterize such mixed sulfates as true chemical compounds or as mixtures; he settled on the compromise term, *mélange chimique*.[44]

Thus, by the end of the second decade of the nineteenth century, the set of discrepancies involving variable composition although the crystal form was fixed, was more prominent than ever, especially because of the production of mixed crystallizations in the laboratory. These had serious implications for Haüy's doctrine of fixed mineral species. Beyond this, they assailed Haüy's molecular theory of crystal structure. In terms of his theory, it was difficult to see how the mixed crystallizations of the type studied by Beudant could be easily explained.

Even more detrimental to Haüy's molecular concept were the cases where apparently identical substances assumed different and incompatible crystal forms. Indeed, Haüy himself had written, in his *Tableau comparatif*, that he could accept the crystalline identity of calcspar and iron spar without any great harm being done to his theory; what would be impossible was that the same substance exhibit incompatible crystal forms.[45] Yet from almost the inception of his theory of crystal structure, there existed a serious anomaly of this type, and one involving that most important of all mineral substances to the crystallographer: calcspar. One could almost write the early history of crystallography—up through Haüy's own initial success in the *Essai* of 1784—in terms of demonstrations that the various shapes assumed by calcspar crystals could be derived from a common rhombohedral form. But by 1788, another mineral, discovered in Spain and named *aragonite* by Werner, had been analyzed by Klaproth, and found to be composed of calcium carbonate, as were calcspar crystals.[46] Haüy, however, found that the primitive form of aragonite was a regular octahedron, a form incompatible with that of calcspar, in that crystals of the one type of mineral could not be related to the primitive form of the other by simple and symmetrical laws of decrement. Moreover, calcspar and aragonite crystals differed from each other in such physical properties as their hardness and density.

Other examples of crystallographical incompatibility allied with apparent identity of chemical composition were subsequently discovered,[47] but calcspar and aragonite remained the most significant case. Those who attacked Haüy's species concept—with its correlation between form and composition—usually cited this anomalous pair of mineral substances as the crucial case of its failure. Others, such as Laplace, suggested that a difference in the arrangement of the *molécules intégrantes* among themselves might account for the difference in crystalline form—without, however, elaborating on how this would work in Haüy's theory.[48] These minerals were subjected to a great deal of chemical analysis, optical study, and crystallographical examination, in the hope that some fundamental chemical difference would be discovered, or that their crystal forms could be reconciled.[49] Although hopes were raised, on both the chemical and crystallographical fronts, that the anomaly was resolvable, they proved to be false, and by the end of the second decade, the calcspar-aragonite anomaly and similar cases also remained as elusive of solution as ever.

These two sets of anomalies to the strict correlation between form and composition were inexplicable as long as one assumed, as Haüy did, that the compound *molécule intégrante* was composed of determinately shaped elementary molecules and that the form of the compound molecule was somehow the sum of the elementary forms. But both these anomalies were resolved by a young German chemist, Eilhard Mitscherlich, when he was just at the start of his career. Mitscherlich was new to both chemistry and crystallography when he began his investigations of these problems, having come to the subject by way of oriental studies and medicine.[50] The factors of novelty may well have been crucial in Mitscherlich's success, for his "resolution" of the anomalies was less the discovery of new phenomena than it was the viewing of long since recognized problems both more comprehensively and in somewhat different terms than had his predecessors and contemporaries. This was particularly true of his first set of studies, those of sub-

---

[44] F. S. Beudant, "Recherches sur les causes qui déterminent les variations des formes cristallines d'une même substance minérale," *ibid.* **3** (1818): p. 254.

[45] R. J. Haüy, *Tableau comparatif*, p. 125. *Cf.* below, chap. 6, for Pasteur's echo here.

[46] *Cf.* H. Hartley, *Polymorphism: An Historical Account* (Oxford, Holywell Press, 1902), p. 8; J. G. Burke, *op. cit.*, p. 126.

[47] Anatase and rutile (titanium oxide), *cf.* L. Vauquelin, "Expériences sur l'anatase qui prouvent que cette substance est un métal," *Jour. des mines* **11** (1801): pp. 425–432; pyrites and marcasites (iron sulfide), *cf.* L. P. Dejussieu, "Sur le fer sulfure blanc," *ibid.* **30** (1811): pp. 241–253.

[48] "Les molécules intégrantes peuvent s'unir par diverses faces, et produire ainsi des cristaux différentes par la forme, la dureté, la pesanteur specifique et leur action sur la lumière," "Considérations générales. 2e supplement aux livre X," *Mécanique céleste. Œuvres complètes de Laplace* (14 v., Paris, Gauthier-Villars, 1878–1912) **4**: p. 490. The context was a discussion of the reduction to physics of material phenomena through the concept of molecular equilibrium, the result of molecular attraction, molecular shape, and caloric repulsion.

[49] *Cf.* J. G. Burke, *op. cit.*, pp. 126–130.

[50] Mitscherlich is in dire need of an extended, scholarly study. *Cf.* R. Winderlich, "Eilhard Mitscherlich (1794–1863): His Life and Achievements," *Jour. Chem. Ed.* **26** (1949): p. 358 for some details on his early career. Also, F. Szabadváry, "Mitscherlich, Eilhard," *Dictionary of Scientific Biography* **9**: pp. 423–426.

stances of different chemical composition but the same crystal form, which Mitscherlich named *isomorphs*.

The investigation of isomorphic substances prior to Mitscherlich had primarily been concerned either with problems of crystal formation and description of crystal structure, or with problems of mineral classification. Mitscherlich, on the other hand, attacked the investigation from the point of view of Daltonian chemistry. He looked for analogies between the weight proportions of the constituents of substances which crystallized in the same form. Salts of the phosphates and the arsenates provided the observational clues to his further studies of the various metallic sulfates. He was struck by the fact that the weight ratios of oxygen in samples of phosphorous and phosphoric acid (with constant weights of phosphorus), and in arsenious and arsenic acid (also with fixed weights of arsenic), was three to five. The corresponding salts of phosphorus and arsenic crystallized in the same form.[51] In his studies of the sulfates, he extended his search for analogous weight proportions even to their water of crystallization.[52] Weight and form, then, were Mitscherlich's guiding parameters, but related now by the arrangement of Dalton's gravimetric atoms, rather than by the shapes of Haüy's elementary molecules themselves. Thus, Mitscherlich defined isomorphism:

... the law of the relation between chemical composition and crystalline form could thus be enunciated: *the same number of atoms combined in the same manner produced the same crystalline form; and [the same crystalline form] is independent of the chemical nature of the atoms and it is only determined by the number and relative position of the atoms.*[53]

Significantly, in this first definition of isomorphism, the chemical nature of the constituents was consciously neglected. Atoms, whatever their chemical nature, of the same number and arrangement, produced analogous chemical formulas and identical crystalline form. Moreover, mixed crystallizations of such isomorphic substances were possible in any proportion. It was only later that Mitscherlich modified his definition to take into account the chemical nature of isomorphous constituents.

Mitscherlich was led to the study of the other major problem—substances of identical composition which crystallized in different forms—from an anomaly in his studies of phosphate-arseniate isomorphs. He found that the crystalline form of the biphosphate of sodium (Na H_2PO_4-H_2O) was *not* isomorphic with the corresponding arsenate salt. While retesting the chemical composition of the biphosphate salt, Mitscherlich produced crystals of it which were isomorphic with the biarsenate sodium salt, but whose form was incompatible with the previously studied biphosphate crystals.[54] What made all of this especially noteworthy was that Mitscherlich himself had prepared these crystals artificially from chemically pure samples of the biphosphate salt. All the previous examples of such dimorphosm, as it was later called, such as calcspar and aragonite, had been found in mineral samples, whose purity was obviously not under the control of the investigators.

Moreover, as with isomorphism, the Daltonian orientation of Mitscherlich gave him flexibility; it freed him from the constraint felt by Haüy, against accepting the possibility of the same substance crystallizing in incompatible forms. Indeed, Mitscherlich thought that the atomic theory provided a ready explanation:

This phenomenon is easy enough to explain in terms of the corpuscular theory. If the relative position of the atoms which have produced a crystal is changed by some circumstance, the primitive form will no longer remain the same.[55]

He considered the case of calcspar and aragonite as being explicable in the same manner as the biphosphate salt, and he thought that this reasoning was strengthened by the fact that calcspar and aragonite were each isomorphic with a different set of crystalline carbonates: calcspar with iron and magnesium carbonate, and aragonite with lead and strontium carbonate.[56]

---

[51] E. Mitscherlich, "Ueber die Krystallisation der Salze in denen das Metall der Basis mit zwei Proportions Sauerstoff verbunden ist," [*Berlin, Abhandl.*, 1818–1819, 427–437] *Gesamalte Schriften*, ed. A. Mitscherlich (Berlin, Ernst Siegfried Mittler und Sohn, 1896), pp. 123–124. Here, he noted that he was struck by the fact that phosphorus and arsenic each combined with five "volumes" of oxygen to form their acids; each acid produced salts, with the same base, whose crystal forms were the same.

In the *Annal. de chimie* redaction, he said he was struck by the fact that the proportions of oxygen to fixed weights of phosphorus or arsenic in the *two* acids of each of these elements, were in the ratio of 3:5. "Sur la relation qui existe entre la forme cristalline et les proportions chimiques; Premier mémoire sur l'identité de la forme cristalline chez plusieurs substances différentes," *Annal de chimie* 14 (1820): p. 172. *Cf. Gesammelte Schriften*, p. 123 for his introductory comments relating his isomorphic concepts to the atomic theory. Here, however, he refused to choose between the atomic theory and the "dynamical" point of view. This introductory section was omitted from the *Annal. de chimie* paper.

[52] *Ibid.*, pp. 129–130.

[53] E. Mitscherlich, "Sur la relation qui existe entre la forme cristalline et les proportions chimiques: IIme mémoire sur les arsenates et les phosphates," *Annal. de chimie* 19 (1821): p. 419. This paper was translated from its original Swedish by Mitscherlich himself for the *Annal. de chimie*. In the introduction, where he first used the term, "isomorphes" [*ibid.*, p. 351], Mitscherlich was much more positive about relating his isomorphic concept to the Daltonian atomic theory than he had been in the previous paper (e.g., he no longer wrote of "volumes" but of atoms). *Ibid.*, p. 350. Also, his account of the first isomorphic substance he had studied was different here; he said that they had been various isomorphic sulfates. *Ibid.*, pp. 350–351.

[54] *Ibid.*, pp. 411–412.

[55] *Ibid.*, p. 416.

[56] *Ibid.*, pp. 417–419. *Cf.* below, chaps. 4 and 6, for Laurent's and Pasteur's attempts to relate isomorphism and polymorphism.

Two years after his first explication of dimorphism in 1823 Mitscherlich discovered that sulfur could assume two different crystalline forms: octahedra with rhombic bases were produced from sulfur dissolved in carbon disulfide, and oblique prisms with rhombic bases from crystallizations of fused sulfur.[57] For the next decade, Mitscherlich led the research in the discovery of more such isomorphic and dimorphic substances and the conditions under which they were produced.

With the exception—an important exception, to be sure—of producing dimorphic substances in a laboratory state of purity, Mitscherlich's resolution of the two problems was less a result of new evidence than of novel orientation towards Daltonian atomism. The most general result of Mitscherlich's work was that the rigid correlation between crystalline form and chemical composition, which Haüy had drawn for his fixed species definition, and which had been based on his conception of the *molécule intégrante,* was sundered once and for all. More particularly, the primacy Haüy had accorded to crystallographic data in species determination—this itself, the result of anomalies which would subsequently be characterized as examples of isomorphism—now seemed to have been based upon an overly enthusiastic conception of the power of crystallography as compared with chemistry, to ascertain "true" chemical composition. Implicitly, at least (for Mitscherlich was not explicit on this point), he very conception of Haüy's *molécule intégrante,* the compound crystalline molecule formed out of determinately shaped elementary molecules, was called into question by Mitscherlich's formulation of isomorphism and dimorphism.

The reaction of Haüy to Mitscherlich's work, particularly to his ideas on isomorphism, was sharp. In 1822 Haüy, now almost in his eightieth year, published two monumental restatements of his life's work: a second edition of the *Traité de minéralogie,* and a massive treatise on crystallography, the *Traité de cristallographie.* In both works, Haüy attempted to defend his traditional position on mineral species, stressing the dependence of it on his conception of the *molécule intégrante* as composed of specifically shaped elementary molecules and even claiming that Mitschericsh's discoveries served to increase the primacy of crystallography over chemistry in mineral species determination.[58]

But Haüy was old and his reaction, though understandable, was hardly typical. More illustrative of the impact of Mitscherlich's ideas was Berzelius. Berzelius had taken an interest in Mitscherlich's work very soon after the latter had produced his first paper on isomorphism. From December, 1819, until 1821, Mitscherlich worked with Berzelius in Stockholm. Berzelius regarded Mitscherlich's discoveries as of fundamental importance to both chemistry and mineralogy. He seized upon the principle of isomorphism as an especially useful aid in the determination of ambiguous chemical formulas and hence, of atomic weights. With regard to mineralogy, the formulations of Mitscherlich caused Berzelius to shift his views on mineral classification away from their earlier accord with Haüy's. Although he had emphasized the role of chemistry in mineral classification in his *Nouveau système de minéralogie,* he had not seen any major discrepancy between the chemical and the crystallographical approach to species determination. However, after Mitscherlich, Berzelius could write, in an article of 1822, on what he had determined were a group of isomorphous minerals:

I consider it as perfectly proved that Haüy's idea that the geometrical form is the most essential character of a mineral species can no longer be exact. One will soon see proofs contrary to this idea of M. Haüy multiplied to infinity.[59]

Isomorphism indeed, also seemed to make more uncertain, for Berzelius, the fixity of species which he had earlier found in the laws of definite and multiple proportions:

It seems then to be demonstrated, first, that in the present state of our knowledge, it is impossible to determine satisfactorily, with regard to minerals in which isomorphous substitution prevails, those that compose mineral species.[60]

Mitscherlich's principles of isomorphism and polymorphism had thus indeed undermined Haüy's own concept of the *molécule intégrante* and its corollary, the species definition. And, outside France, where Haüy's molecular crystallography had never been without critics and his approach to mineralogy never without competitors, Mitscherlich may well have been viewed as triumphing over the aged, and now outdated, crystallographer.

But in France, the dominance of Haüy's achievements was too strong to be so easily destroyed. His

---

[57] E. Mitscherlich, "Ueber des Verhaltnis der Krystallformes zu den chemischen Proportionen," [*Berlin, Abhandl.,* 1822–1823: pp. 25–41, 43–48] *Gesammelte Schriften,* pp. 190–194.

[58] R. J. Haüy, *Traité de minéralogie* 1 (1822): pp. xiii–xvii, 37–49, 486–487; *Traité de cristallographie* 2: pp. 409 ff., where Mitscherlich was not, however, mentioned.

[59] J. J. Berzelius, "Analyses de substances minérales (extrait de journaux): Sur les amphiboles," *Annal. des mines* 7 (1822): p. 225. For a more general treatment of this question, cf. "Berzelius über Mineral systeme," *Schweigger, Jour.* [*Jour. f. Chem. u. Phys.*] 36 (1822): pp. 414–422 [this journal simultaneously entitled *Jahr. d. Chem. u. Phys.* 6.].

[60] J. J. Berzelius, "On the Alterations That Must be Made in the System of Chemical Mineralogy in Consequence of the Property of Isomorphous Bodies to Replace One Another in Indefinite Proportions," *Ann. Phil.* 11 (1826): p. 385. Berzelius also changed his basis of classification of species for higher divisions, etc., from the electropositive elements, in which isomorphic substitution was most frequent, to the electronegative ones where, at the time, fewer cases were found. *Ibid.,* pp. 385–386, 422–434.

crystallography and mineralogy continued to be taught. Moreover, there had already appeared by the time of Mitscherlich's first work on isomorphism, a refinement to Haüy's molecular concept—or rather, an attempt to combine it with the chemical atomism of Dalton—which offered the potential not only for reconciling Mitscherlich's principles with Haüy's molecular concept, but also for incorporating them into it. It is to this attempt to construct a "geometrical chemistry" out of Haüy's molecules and Dalton's atoms and the development out of it of what I shall call a crystallographical-chemical approach that we shall now turn.

IV. MOLECULAR GEOMETRY

So far in this account of the development of Haüy's theory of molecular crystal structure and mineral taxonomy, the *molécule intégrante* has been treated as a static, fully formed unit. For neither in his theory of crystal structure nor in his mineralogy, had Haüy been particularly interested in speculating about the internal structure of his *molécules intégrantes* or in their chemical mode of formation.[1]

But after the appearance of Dalton's chemical atomic theory, interest in speculating about the actual arrangement of atoms in molecules did develop among some scientists, especially in France. These scientists attempted, in one way or another, to bring together the theory and data of Haüy's crystallography and Daltonian chemical atomism. Haüy's polyhedral *molécule intégrante* remained the root concept, but it was refined by the introduction into it of Daltonian atoms to produce models of molecular systems composed of atoms arranged symmetrically in space. This type of speculation seems to have begun with an attempt by the physicist André Marie Ampère to work out a general model for chemical combination in geometrical terms, a sketch of which he published in 1814.[2]

The speculative nature of much of this activity must be emphasized; there was no real proof about the actual arrangement of atoms in space until the twentieth century, nor any very solid grounds for such speculation until the appearance of the valence concept in the 1850's. Nor did it preoccupy many chemists or crystallographers in the first half of the century. Yet it was an important development, issuing by mid-century directly or indirectly in the advances in the theory of crystal structure made by Gabriel Delafosse and Auguste Bravais as well as in the discoveries of Louis Pasteur on the correlation between enantiomorphism, isomerism, and optical activity for tartrate salts: watersheds in the history of theories of material structure.

The French approach to atomic arrangement and crystal structure was not the only one adopted to elucidate these issues in the light of Dalton's atomic theory. There was an alternative approach, developed particularly in England by one of Dalton's early supporters, William Hyde Wollaston, and based on the packing together of spherical and spheroidal atomic elements to form crystals.

Dalton himself did not indulge in anything very elaborate in the way of speculations about crystal structure. Yet, the little attention he did bestow on it was significant. In the first volume of *A New System of Chemical Philosophy,* Dalton gave brief attention to this subject and suggested how different regular crystalline shapes, such as cubes and tetrahedra, could be built up by the regular stacking of atomic spheres. Although accompanied by diagrams similar to those used in the seventeenth century by Robert Hooke, Dalton's discussion occupied but a paragraph. However, it was a strategically placed one, since it came immediately before his account of atomic stoichiometrics. He ended his discussion of crystal structure and introduced his thought about chemical atomism with prudent, but prophetic words:

Perhaps, in due time, we may be enabled to ascertain the number and order of elementary particles, constituting any given compound element, and from that determine the figure which it will prefer on crystallization, and *vice versa;* but it seems premature to form any theory on this subject, till we have discovered from other principles the number and order of the primary elements which combine to form some of the compound elements of most frequent occurrence; the method for which we shall endeavor to point out in the ensuing chapter.[3]

Speculation about crystal structure was taken up almost immediately by one of Dalton's first supporters, the polymath scientist, William Hyde Wollaston. Wollaston was conversant with both crystallography and chemistry, to each of which he made important contributions. In the former field, he had invented the reflective goniometer, an improvement over the earlier contact goniometer in the precision of its measurements of crystalline interfacial angles. In chemistry, among other things, Wollaston provided, along with Thomas Thomson and almost simultaneously, the first experimental corroboration of Dalton's law of multiple proportions.[4] Given his dual interests and accomplish-

---

[1] *Cf. Traité de crystallographie* 2: pp. 429–430 and *Atlas,* pl. 69, fig. 13 and 14, where Haüy argued that different mineral species can have the same (cubic) *molécule intégrante* and illustrated this with two square "mosaics" composed, one of a square and four right triangles, and the other of a rhomboid and four trapezoids. This was hardly a serious attempt.

[2] A. M. Ampère, "Lettre de M. Ampère a M. le Comte Berthollet, sur la détermination des proportions dans lesquelles les corps se combinent d'après le nombre et la disposition respective des molécules dont leurs particules intégrantes sont composées," *Annal. de chimie* 90 (1814): pp. 43–86. A shorter version of this chapter was published by me as "The Atomic Structural Theories of Ampère and Gaudin: Molecular Speculation and Avogadro's Hypothesis," *Isis* 60 (1969): pp. 61–74.

[3] J. Dalton, *A New System of Chemical Philosophy* 1: p. 211.
[4] D. C. Goodman, "Problems in Crystallography in the Early

ments, Wollaston was particularly well placed to suggest a new synthetical approach to the theory of crystal structure. In the paper in which he verified the law of multiple proportions, Wollaston had drawn a conclusion which, as it were, turned round Dalton's precautions against speculating about crystal structure, into a prescription for future investigation:

> ... and I am further inclined to think, that when our views are sufficiently extended, to enable us to reason with precision concerning the proportions of elementary atoms, we shall find the arithmetical relation alone will not be sufficient to explain their mutual action, and that we shall be obliged to acquire a geometrical conception of their relative arrangement in all three dimensions of solid extension.[5]

Four years later, in his Bakerian Lecture of 1812, Wollaston himself returned to his proposal of 1808 to elaborate on it, but his focus had shifted: from a chemical orientation—the union of atoms into molecules—to a more physical one—crystal structure itself. He proceeded to mount a critique of Haüy's molecular theory of crystal structure and to propose an alternative approach of his own. His point of departure was the case of the structure of fluorspar, a serious anomaly in Haüy's theory from its inception. The form of the fluorspar molecule could either be tetrahedral or octahedral in Haüy's theory; there was no way to decide which. More seriously, in either case, vacua between the molecules of a fluorspar crystal were necessitated.

Wollaston seized on this last point to argue that the structural stability of fluorspar crystals, composed of molecules themselves in contact only along their edges and interspersed with vacua, would be precarious. A more stable structure would obtain in a model based upon the close packing of spheres or spheroids to make up the crystal structure.[6] Wollaston had argued similarly from macromechanics to micromechanics in his 1808 paper. Thus, with reference to molecules composed of three atoms of one substance to one atom of another, or of four atoms to one atom, he had written then:

> If there be three [to one], they might be arranged with regularity, at the angles of an equilateral triangle in a great circle surrounding the single spherule; but in this arrangement, for want of similar matter at the poles of this circle, the equilibrium would be unstable, and would be liable to be deranged by the slightest force of adjacent combinations; but when the number of one set of particles exceeds in the proportion of four to one, then, on the contrary, a stable equilibrium may again take place, if the four particles are situated at the angles of the four equilateral triangles composing a regular tetrahedron.[7]

It was on this criterion of stability that Wollaston based his 1812 alternative to Haüy's theory of crystal structure. He proceeded to demonstrate how almost all of the crystalline forms could be generated by the close packing of spherical or spheroidal elements. Wollaston was quite aware of his seventeenth-century predecessors, Hooke and Huyghens, but he pointed out that neither of them had been concerned to work out crystal structure generally and systematically.[8]

In his 1812 lecture, Wollaston had moved somewhat away from his earlier post-Daltonian chemical orientation; his spheres and spheroids were now essentially physical rather than chemical elements. However, in his considerations of the one serious anomaly to his own model of crystal structure, Wollaston was brought, in recourse, back to Daltonian chemical proportions. Paradoxically, this anomaly was his geometrically simplest case: the cube. The problem was mechanical stability. It was simple enough to imagine four atop four spheres aggregated to form a cube; Dalton himself had made this obvious suggestion. But Wollaston realized that neither this, nor any other juxtaposition of homogeneous spheres would yield the requisite structural stability. One could imagine a cube built up out of oblate spheroids to yield what Wollaston called a "right angled rhomboid." But, though geometrically feasible, it was at variance with the three-dimensional isotropism possessed by most cubic crystals.

There was one other possibility which Wollaston considered: a cube could be built up out of geometrically identical spheres, if the additional assumption were made that the spheres were non-homogeneous in another respect—that they be of two different varieties (pictured by him as black balls and white balls). If one imagined these to be perfectly intermixed, then both the cubic form and the requisite stability (with additional implicit assumptions about chemical equilibria) would be obtained.[9]

Wollaston was aware of the chemical significance of his cubic model: that it not only satisfied the require-

---

Nineteenth Century," *Ambix* 16 (1969): pp. 152–166; "Wollaston and the Atomic Theory of Dalton," *Historical Studies in the Physical Sciences* 1 (1969): pp. 37–59.

[5] W. H. Wollaston, "On Super-Acid and Sub-Acid Salts," *Phil. Trans.* 98 (1808): p. 101. He did qualify this with: "It is perhaps too much to hope, that the geometrical arrangement of primary particles will ever be perfectly known...." *Ibid.*, p. 102.

[6] W. H. Wollaston, "The Bakerian Lecture: On the Elementary Particles of Certain Crystals," *Phil. Trans.* 103 (1813): pp. 51–63.

[7] W. H. Wollaston, *Phil. Trans.* 98 (1808): p. 102.

[8] W. H. Wollaston, *Phil. Trans.* 103 (1813): pp. 53–57. He claimed to have been referred to Robert Hooke's speculations on crystal structure in the *Micrographia* through "the kindness of a friend," *ibid.*, p. 53. He unquestionably would have known of Huygens's model from his own earlier verification of Huygens's construction for double refraction. *Cf.*, "On the Oblique Refraction of Iceland Crystals," *Phil. Trans.* 92 (1802): especially p. 382.

[9] W. H. Wollaston, *Phil. Trans.* 103 (1813): pp. 59–62. He cited as a success of this the ready explanation of the electrical polarity of boracite. *Ibid.*, p. 61.

ments for stable structure but also dovetailed, in its one-to-one ratio of different types of atoms, with Daltonian stoichiometry:

> An hypothesis of uniform intermixture of particle with particle, accords so well with the most recent views of binary combination in chemistry, that there can be no necessity, on the present occasion, to enter into any defence of that doctrine, as applied to this subject.[10]

Though Wollaston himself developed these ideas no further, his Bakerian Lecture of 1812 set the tone for speculation on crystal structure in England and America for the rest of the nineteenth century, through J. W. F. Daniell, James Dwight Dana, William Foster on to the work of Kelvin and William Barlow in the last quarter of the century,[11] and was destined to play a notable role in twentieth-century ideas on material structure. Wollaston had rejected Haüy's theory of crystal structure, with its polyhedral molecular element, in favor of an alternative, based on spherical and spheroidal material units, harking back to Hooke and Huyghens. His successors followed suit in developing this alternative model of crystal structure.[12] They also followed him—at least the Wollaston of the Bakerian Lecture—in focusing more upon the physical rather than the chemical aspects of crystal structure.

If Wollaston fostered an Anglo-American tradition alternative to Haüy's theory of molecular crystal structure, Ampère's paper of 1814 seems to have been the major source for a French tradition in which there was a much more direct synthesis of Haüy's molecular theory with Daltonian atomism. The opening section of Ampère's paper of 1814 has been well known, indeed overshadowing the bulk of the paper by its significance. For Ampère began the paper with an independent restatement of the gas hypothesis first enunciated by Amedeo Avogadro in 1811: that equal volumes of gases contain the same number of molecules at constant pressure and temperature.[13] However, the greater part of the paper, which was dedicated to Berthollet, dealt with quite another, though by no means unrelated matter: Ampère's attempt to elucidate questions of chemical union and chemical proportions through a geometrical theory of chemical combination. His mode of explanation was by means of analogies derived from Haüy's theory of crystal structure but applied to domains of material states on the whole far removed from those to which crystallography applied. Thus, unlike Wollaston, Ampère was not primarily concerned with refining a model for crystal structure *per se;* it was rather the application of crystallographical concepts to chemistry—particularly to post-Daltonian stoichiometry —that interested him.

Although he is best known for his work in electrodynamics and mathematics, Ampère also took an avid interest in chemistry. He taught it in his first teaching position in Lyon and Bourg, and he continued to follow developments closely in this science after coming to the *École Polytechnique* as a mathematician in 1803, even performing his own chemical experiments. He was in correspondence with Humphry Davy, to whom he was eager to communicate his atomic speculations.[14]

The roots of Ampère's knowledge of crystallography are less clear, but he certainly derived essential ingredients for his own speculations from Haüy's theory of molecular structure: specifically, he took over the notion of the polyhedral *molécule intégrante* from Haüy, but he introduced into it the chemical atom. He endeavored to work out models of elementary atoms arranged in chemical union to form polyhedral compound molecules.

What he hoped to do with his models was to ascertain unambiguously the known combining proportions of chemical substances and to predict other ones. He himself stated what advantages he though his "geometrical" method had over simple Daltonian stoichiometrics:

> . . . but when one makes use of it [the law of multiple proportions], nothing indicates how great a proportion of an elementary substance ought to enter into one of its compounds: whereas the consideration of the representative forms enables one to predict, in many cases, how the molecules of each of its elements ought to enter [into combination] in a compound substance and even leads one to establish, between the combinations of two elementary substances with all the others, a dependence such that, when the combinations of one of these substances is known, one can predict those of the other.[15]

As the starting point for his molecular geometry, Ampère adapted one of Haüy's crystallographical concepts: the "primitive form." This, it will be recalled, was the form of the crystalline nucleus, often revealable by cleavage and alike in crystals of the same mineral species. Haüy had originally identified the form of the macroscopic nucleus with that of its component *molécules intégrantes,* but he had eventually come to distinguish between nuclear and molecular forms, even though they remained closely related in his molecular theory of crystal structure. Ampère, however, utilized five of the six shapes Haüy had designated as those

---

[10] *Ibid.*

[11] This "Anglo-American" tradition, if it may be called that, is itself in need of study. *Cf.* the comments of L. Sohncke in his *Entwickelung einer Theorie der Krystallstruktur* (Leipzig, B. G. Teubner, 1879), pp. 14–18.

[12] *Cf.* J. F. Daniell, "On the Relation Between the Polyhedral and Spheroidal Theories of Crystallization; and the Connexion of the Latter with the Experiments of Professor Mitscherlich," *Roy. Inst. Jour.* **2** (1831): pp. 30–44.

[13] A. M. Ampère, *op. cit.*, pp. 46–47.

[14] *Cf.* L. de Launay, *Le grand Ampère d'après des documents inédits* (2nd ed., Paris, Perrin, 1925), especially pp. 158 ff. Also, L. Pearce Williams, "Ampère, André Marie," *Dictionary of Scientific Biography* **1**: pp. 142–143.

[15] A. M. Ampère, *op. cit.*, p. 73.

of the primitive forms as the simplest ones of the molecules themselves.[16] (Ampère, using a terminology derived probably from Fourcroy,[17] termed *molécules intégrantes* "*particules*"; and the constituent atoms, "*molécules*." We shall follow Ampère's terminology while dealing with his theory.) Moreover, he went beyond anything in Haüy's theory by trying to work out the structure of his *particules*. These were envisioned as made up of *molécules* (atoms) fixed at the summits of the solid angles; hence the geometrically simplest *particule* for Ampère—the regular tetrahedron—was composed of four *molécules*.

In addition, Ampère meant to encompass in his theory not just crystal structure, or even particularly this domain, but the molecular structure of matter in all its physical states, with particular attention given to the gaseous state. The recently discovered data concerning this latter, such as that on combining volumes of gases, seemed to provide clues to the forms of the gaseous *particules*, both elementary and compound, and hence, to the numbers of *molécules* they contained. With this as a basis, Ampère believed that one could proceed to work out the *particule* form eventually of any substance and indeed to come full circle—to convert the "conjectures" about the gas law itself into certain knowledge.

The mediating assumption by which *particule* shapes in all physical states were related was that *particules* retained their fundamental polyhedral form in any state. His statement and utilization of the gas hypothesis, for which the paper is famous, was based upon this assumption:

I have proceeded ... from the supposition that, in the case where bodies pass to the gaseous state, their *particules* alone are separated and distanced one from another by the expansive force of caloric, to distances very much greater than those at which the forces of affinity and of cohesion have an appreciable action, such that these distances depend only on the temperature and pressure which supports the gas and that at equal temperatures and pressures, the *particules* of all gases, be they simple or compound, are placed at the same distance from one another. The number of *particules* is in this supposition proportional to the volume of the gas.[18]

In hindsight, there was a beautiful irony in Ampère's extension of the crystallographical molecular concept to gases. For twentieth-century crystallographical analysis has shown that the source of his molecular concept—the domain of inorganic crystal structure—was the principal one where this molecular concept least applies; in such crystal lattice structure, discrete physical molecular units, like Haüy's *molécules intégrantes*, usually do not exist as such.[19] On the other hand, it is the gaseous domain, where the molecular concept—premising the existence of discrete molecules—is the most applicable.

But to return to Ampère's molecular geometry for the chemical combination of gases: As "the most simple supposition" about gaseous *particule* shape, Ampère postulated a four-*molécule* tetrahedron for the elemental gases oxygen, nitrogen, and hydrogen.[20] Employing the gas hypothesis, he then worked out models of chemical combination between gases. Since all his *particules*, including the gaseous ones, were polyhedra containing at least four *molécules* each, Ampère was able to explain the combining volume data of Gay-Lussac, which so puzzled all his contemporaries except Avogadro, by assuming that these *particules* could split up and the component *molécules* recombine to form new polyhedral *particules* with the *molécules* of other substances. For example, one volume of nitric oxide gas (NO, *gaz nitreux*) contained one-half volume of oxygen. In Ampère's terms, this meant that nitric oxide gas was made of a half of a *particule* of nitrogen (two out of its four *molécules*) which had united with a half of a *particule* (again, two *molécules*) of oxygen to form one compound *particule* (four *molécules*) of nitric oxide.

Analogous reasoning suggested that *particules* of nitrous oxide (*oxide d'azote*) were made of six *molécules*: four of nitrogen and two of oxygen; *particules* of water were composed of four *molécules* of hydrogen united to two of oxygen; *particules* of ammonia contained eight *molécules*: six of hydrogen and two of nitrogen. Four, six, and eight *molécules* composed respectively tetrahedral, octahedral, and parallelepipedal *particules*—three of the five primitive forms which Ampère thought simple *particules* assumed.[21]

More generally, Ampère attempted to work out models of chemical union, which he defined geometrically as the mutual interpenetration of two or more *particules* until their centers of gravity were coincident. He determined the shapes of a large number of complex polyhedra which he thought would be produced by such interpenetration. The regularity or lack thereof of the resultant compound *particule* gave Ampère what he took to be a criterion for the feasibility of a chemical reaction. For example, if the *particules* of one of two combining substances were tetrahedral, and those of the other were octahedral, two of the latter could not unite with only one of the former to form a compound, for they would be unable to produce a regular polyhedron.[22]

---

[16] Parallelepiped, regular tetrahedron, octahedron with triangular faces, hexagonal prism, rhombic dodecahedron. *Ibid.*, pp. 49–50.

[17] A. F. Fourcroy, *Philosophie chimique* (3rd ed., Paris, Bernard, 1806), pp. 42–43.

[18] *Ibid.*, pp. 46–47.

[19] *Cf.* C. S. Smith, "Structural Hierarchies in Inorganic Systems," *Hierarchical Structures*, ed. L. L. Whyte, A. G. Wilson, D. Wilson (New York, American Elsevier, 1969), pp. 61–63.

[20] A. M. Ampère, *op. cit.*, p. 48.

[21] *Ibid.*, pp. 47–49.

[22] *Ibid.*, pp. 56–57.

Ampère suggested the wide range of application his chemical geometry might have: for example, to the stoichiometry of neutral salts as compared with acidic and basic salts of the same base, to the proportions of water of crystallization (which he regarded as an integral chemical constituent) in crystalline salts, and even to the proportions of gases which could dissolve in a given quantity of water.[23] He also took up some of the main problems and novelties of the day, especially those of Davy, as examples of areas of applicability. These included the proportions of oxygen which could unite with potassium in the latter's various physical states, the question of the proportions of nitrogen and oxygen in their various oxides and acids, and the chemical composition of the oxychloride compound Davy had dubbed "euchlorine." Ampère's strategy, particularly in the last two examples, was to use the volumetric proportions of the initial gaseous constituents, and the volume of the final product to construct the form of the new compound's *particules* out of the constituent elementary *particules,* about whose shapes he had already made assumptions.[24]

In one example, ammonium chloride, Ampère referred to crystallography for corroboration. Like most of his cases, his two constituents here were gases. Ampère thought that chlorine *particules* contained eight *molécules;* therefore, the hydrochloric acid *particule,* composed of half of a *particule* each of chlorine and hydrogen, had the form of an octahedron, made up of four chlorine and two hydrogen *molécules.* Now, if this be united with a cubic (for simplicity [!]) *particule* of ammonia, a sal-ammoniac *particule* in the form of a rhombic dodecahedron would result. Since sal-ammoniac crystallized in the same system of symmetry as the rhombic dodecahedron (the cubic system) so that the real external shapes of sal-ammoniac crystals were geometrically relatable to rhombic dodecahedra, Ampère felt he had "proof" of his reasoning.[25]

Ampère's paper of 1814, though only a sketch of his theory, was the most extensive account of it that he published. He continued to think it an important subject and he lectured on it at the Collège de France, but his only published return to the subject and to the gas hypothesis was a brief one nearly two decades later, in a paper of 1832 which dealt primarily with the nature of light and heat. Here, Ampère not only reasserted the gas hypothesis, but also distinguished clearly between "atoms" and "molecules" (which he now denoted by these terms). Moreover, he still retained his crystallographically inspired mode of conceiving molecules:

It follows from this definition of molecules and of atoms that the molecule is essentially solid, that the bodies to which they pertain are solid, liquid or gaseous; that molecules necessarily have a polyhedral form of which their atoms, or at least a certain number of their atoms, occupy the apices; and that these polyhedral forms are those which are designated under the name of primitive forms by the crystallographers.[26]

Ampère's program for a geometrical chemistry was ingenious and imaginative, but by the same token, it was also highly speculative and even fanciful. There was much in the 1814 paper which was arbitrary and *ad hoc*. Perhaps for this reason Ampère's ideas and general program were not developed immediately. Yet, by the time of the appearance of Ampère's second paper, his general approach was about to be taken up by members of a new generation of French chemists: specifically, by Marc Antoine Gaudin, Alexandre Eduard Baudrimont, and Auguste Laurent. We might pause to explain what is meant by Ampère's "general approach" especially since each of these three chemists followed his own non-conformist and independent pursuit. There is little evidence of any particular friendship between these three men, though they each knew, and in some cases were influenced, by the others' ideas. Nor did they all equally acknowledge Ampère's paternity; they ranged from Gaudin who claimed to be Ampère's disciple, to Laurent who gave Ampère no credit in this connection, though it is striking that all three subscribed to the gas hypothesis which they credited to Ampère.

What they had in common with each other—and with the Ampère of 1814—was their view that the key to understanding chemical phenomena was a geometrical one; that this understanding lay in the explication of the arrangement of atoms into polyhedral molecules; and that the study of crystal form had to be coordinated with that of chemical composition in order to get at molecular form and atomic arrangement. By the early 1830's, new data and even new domains of chemistry had opened up the possibilities of developing this approach beyond Ampère's 1814 program, coming as it had at virtually the birth of modern quantitative chemistry: new, and apparently anomalous, vapor density data compiled in the 1800's, the appearance of new concepts relating chemistry and crystallography, such as Mitscherlich's isomorphism and polymorphism, as well as isomerism (with its strong implications for the significance of atomic arrangement) and finally, the emergence of organic chemistry.

The first of these developments, the extension of the gas law to new vapor-density data, was the preoccupation of Gaudin. In 1831 he submitted a paper to the Académie des Sciences on the gas hypothesis, in which he took up the general consideration of the gas hy-

---

[23] *Ibid.*, pp. 75–76.
[24] *Ibid.*, pp. 76–86.
[25] *Ibid.*, pp. 81–82.

[26] A. M. Ampère, "Note de M. Ampère sur la chaleur et sur la lumière considerées comme resultant de mouvemens vibratoires," *Annal. de chimie* **58** (1835): p. 434.

pothesis and gave what was perhaps the clearest exposition of it between its first enunciations by Avogadro and Ampère on the one hand, and its dramatic utilization by Stanislao Cannizzaro in the late 1850's, on the other.[27] Gaudin was a strange figure, always hovering in the wings of recognition, but never really achieving any scientific renown, despite the fact that his scientific activities were wide-ranging and persisted over a long period of time. His early education and career are obscure, but, from 1835, he was employed at the Bureau des Longitudes in Paris. His lifetime interests included photography as well as mineralogy and chemistry.

The topic to which Gaudin devoted the most attention was that of atomic arrangement. He had been sparked—or so he said—by Ampère's lectures on this subject at the Collège de France in 1827 to develop his own ideas.[28] Despite the lack of evidence of personal interaction between the renowned physicist and his self-proclaimed disciple, Gaudin left no doubt that he owed his inspiration to Ampère. It was from Ampère also that Gaudin claimed to have learned of the gas hypothesis, the subject of the 1831 paper. In this paper, Gaudin set himself the ambitious task of giving a consistent explanation for all the available germane data.

In order to illuminate the difficulties in this task, it is necessary to know the background of Gaudin's considerations of vapor-density data in the experiments of the chemist, Jean Baptiste Dumas and his interpretation of their results. By the early 1830's, Dumas had established himself as unquestionably the leading young chemist in France, having already published several volumes of his textbook, produced, either alone or in collaboration, important work in organic chemistry and compiled the major share of new data on vapor densities.[29]

Dumas was a friend of Ampère, and perhaps played something of the role of intermediary in transmitting Ampère's chemical ideas to his contemporaries and students.[30] His own attitude towards Ampère's chemical geometry was complex and ambiguous. In the 1820's, he seems to have been favorably disposed to it. Subsequently, in the 1830's, he turned sour on it, only still later, after 1838, to reassert his interest in the geometrical approach and eventually to express his keen admiration for Ampère's molecular geometry.[31] Dumas himself does not seem to have indulged in the sort of crystallographical-chemical speculation fostered by Ampère and favored by Gaudin, Baudrimont, and Laurent; on the contrary, as we shall see, he even censured some of those mentioned above for doing so in the 1830's.

In 1826, very early in his career, Dumas published the first of a number of papers, devoted to the determination of correct atomic weights through the measurement of the densities of elementary gases by improved experimental techniques of Dumas's own devising.[32] The inspiration for this work seems to have been Ampère's gas hypothesis. In the 1826 paper, Dumas gave one of the relatively few clear statements of it in this period: ". . . in all elastic fluids under the same conditions, the molecules are found to be placed at equal distances [from each other], that is to say, they are in the same number [for the same volume]."[33] But Dumas himself did not succeed in realizing the full potential of this hypothesis, even though it was his guide in the gaseous density experiments. The reason seems to lie both in his interpretation of the gas hypothesis—an interpretation which differed significantly from Ampère's—and in some anomalous experimental findings which Dumas was unable to square to his own satisfaction with the gas hypothesis.

With regard to his interpretation: while Dumas initially subscribed to the gas hypothesis, he was unwilling to follow Ampère into the speculative maze of atomic

---

[27] M. A. Gaudin, "Recherches sur la structure intime des corps inorganiques," *Annal. de chimie* 52 (1833): pp. 113–133. For a more general account of his speculations, *cf.* "Note sur quelques propriétés des atomes," *Bibl. univ.* 52 (1833): pp. 131–141 (dated 30 Oct., 1831) with introductory comments entitled: "Quelques réflexions sur la théorie des atomes en chimie," (pp. 127–131) by the editor "A. D." (August De la Rive). *Cf.* also a review of Gaudin's ideas by A. C. Becquerel and an account of them by Gaudin himself in *Revue encyclop.*: "Recherches sur les atomes" (Becquerel) and "Resumé de la théorie atomique de M. Gaudin" 56 (1832): 320–329; 329–335 respectively. Both have good historical accounts. For the history of the vicissitudes of this paper, *cf.* M. Delepine, "Une étape de la notion d'atomes et de molécules," *Bull. Soc. chim. France* 2 (1935): pp. 12–13.

[28] The year 1827 is given by C. Graebe, "Der Entwicklungsgang der Avogadroschen Theorie," *Jour. prakt. Chem.* 87 (1913): p. 160. Gaudin wrote of the inspiration of Ampère's *lectures* for him in his *L'architecture du monde des atomes* (Paris, Gauthier-Villars, 1873), pp. xii–xiii.

[29] *Cf.* S. C. Kapoor, "Dumas, Jean-Baptiste-André," *Dictionary of Scientific Biography* 4: pp. 242–244; "Dumas and Organic Classification," *Ambix* 16 (1969): pp. 1–65.

[30] This idea suggested by T. H. Levere, *Affinity and Matter* (Oxford, Clarendon Press, 1971), p. 170, although there is admittedly, little evidence.

[31] S. C. Kapoor, *Ambix* 16 (1969): pp. 7–13. *Cf.* J. B. Dumas, "Remarques sur l'affinité," *Annal. de chimie* 15 (1868): pp. 89–90, where he wrote that the "système moléculaire proposé par Ampère, modifié par M. Gaudin et généralement adopté [!?] avec diverses variantes par les chimistes que, s'occupant de chimie organique, sont obligés de tenir compte des phénomenes de substitution, ait rendu à la fois moins ardente la pour suite d'une théorie electro-chimique précise et moins confiante l'interprétation trop absolu de la nomenclature française." I thank Dr. Levere for this reference. Dumas went on to say: "On se trouve ramené en même temps vers la pensée qui attribue aux molécules de corps composés une constitution plus complex que celle qui deriverait de la nomenclature binaire, et qui en fait des systèmes planétaires ou cristallographiques, offrant la réunion de plusieurs atomes ou centres de force, mobiles dans le premier cas, fixes dans le second." *Ibid.*, p. 90.

[32] J. B. Dumas, "Mémoire sur quelques points de la théorie atomistique," *Annal. de chimie* 33 (1826): pp. 337–391.

[33] *Ibid.*, p. 338.

arrangement. This reluctance on Dumas's part came out clearly in his attempts to grapple with the question of molecular division, required by the coordination of the gas hypothesis with combining volume data. Dumas was cautious in his reasoning from gaseous density and combining volume data to the actual numbers of atoms in gaseous molecules. From the first, he was of the opinion that the existence of polyatomic elementary gaseous molecules, whose existence he always admitted, meant that one was unable actually to get at the numbers of atoms in equal volumes of gases.[34]

This in itself would not have been overly serious had the application of the gas hypothesis to the determination of atomic weights remained restricted to the (diatomic) permanent elementary gases: oxygen, nitrogen, hydrogen, and chlorine. But Dumas extended his vapor density determinations to other substances not usually gaseous—sulfur, phosphorus, arsenic, mercury, and some of their compounds—with much more confusing results, especially given his inhibitions against speculation about the numbers of atoms in gaseous molecules. The confusion on Dumas's part was also aided and abetted somewhat by his failure to keep absolutely clear and uppermost in his reasoning the distinction between the physical gaseous molecule and the atoms which composed it, so as always to be careful of whether he was speaking of equal numbers of atoms or equal numbers of molecules in equal volumes of gases.

It was to the vapor density data of Dumas that Gaudin addressed himself. His relations with Dumas are, like most facts about his personal and professional life at this period, obscure. However, late in his life, Gaudin singled Dumas out among chemists for having supported him and encouraged him to continue his researches; Dumas in his turn (also late in life) gave Gaudin's atomic speculations very high praise. Whether, given Dumas's strictures against such speculations in the 1830's, there was this same mutual regard then, is problematical.

At any rate, Gaudin was able to provide a consistent explanation of all the available data, but an explanation which required the adaptation of speculative hypotheses about the atomic makeup of gaseous molecules. Assuming that such elementary gases as hydrogen and oxygen were diatomic, Gaudin accounted for the seemingly anomalous data for the non-permanent gases by suggesting that their gaseous molecules contained different numbers of atoms from those of the elementary permanent gases. For example, molecules of mercury were monatomic; those of sulfur were hexatomic.[35] Atomic formulas were also worked out for various compounds of arsenic, phosphorus, silicon, tin, titanium, and boron, including the correct formula for silica: SiO_2.[36]

The clarity of Gaudin's treatment of these complex and confusing data hinged, in large part, on his consciousness of the primary requirement to distinguish clearly between "atoms" and "molecules." This he undoubtedly owed to Ampère. Also derived from Ampère was the more general project Gaudin adumbrated in his published papers of this period on the gas hypothesis, of working out atomic arrangements and molecular shapes through the coordination of chemical and crystallographical data.[37] He had, in fact, developed his own ideas on this subject, which, however, received only scant publication at this time. Nevertheless, his ideas on atomic arrangement did attract the favorable attention of the physicist A. C. Becquerel.[38]

Despite Becquerel's notice, Gaudin did not amplify his ideas in print again for almost fifteen years, until 1847 when he commenced publishing a series of articles on atomic arrangements and molecular shapes. In the first of these, he revealed how conscious he was of the tradition in which he was working:

Haüy, by his admirable work, founded mineralogy. It remained to show that cleavages and decrements are powerless, in certain cases, to reach the primitive molecule and above all, the intimate arrangement of its atoms: this is the problem so happily opened up by the genius of Ampère and it is through his profound lessons at the Collège de France that I have made resolution to continue his work.[39]

During the 1850's and 1860's, Gaudin submitted many

---

[34] Dumas had clearly distinguished between "molécules physiques," polyatomic molecules, and "molécules chimiques" or atoms, but in the first volume of his *Traité de chimie appliqué aux arts* (8 v., Paris, Bechet, 1828–1845), he wrote that: "D'après l'ensemble des phénomènes connus, il est évident qu'on ne peut arriver à connaître précisément la valeur de cette molécule chimique; il faut donc se contenter de la molécule physique donnée par les gaz," 1 (1828): p. xxxix. The anomalous vapor densities of sulfur, phosphorus, arsenic, and mercury only strengthened Dumas in these views. *Cf. Leçons sur la philosophie chimique* (Paris, Bechet Jeune, 1837), pp. 266–268.

[35] M. A. Gaudin, *Annal. de chimie* 52 (1833): p. 123 (mercury), sulfur only implicit in this memoir (p. 121), and table of atomic weights, p. 132.

[36] *Ibid.*, pp. 120–131. *Cf.* also *Bibl. univ.* 52 (1833): pp. 136–141, where Gaudin also outlined an astonishing sketch of a plan to use his reformed atomic weights to order the elements into a proto-periodic table—in which even their physical and chemical properties could be predicted. Theron M. Cole, Jr., "Early Atomic Speculations of Marc Antoine Gaudin; Avogadro's Hypothesis and the Periodic System," *Isis* 66 (1975): pp. 334–360.

[37] M. A. Gaudin, *Annal. de chimie* 52 (1833): pp. 113–114.

[38] Republished later by Gaudin in *L'architecture du monde des atomes*, pp. 218–228. Becquerel also gave Gaudin's speculations (as well as Ampère's and Baudrimont's, treating them all as members of something like a tradition) favorable mention in his *Traité de physique considérée dans ses rapports avec la chimie et les sciences naturelles* (2 v., Paris, F. Didot, 1842–1844) 1: pp. 262–263.

[39] M. A. Gaudin, "Recherches sur les causes les plus intimes des formes cristallines," Paris, *Comptes rendus* 25 (1847): p. 665.

papers to the Académie des Sciences in which he elaborated his theory, and which were duly synopsized in the *Comptes rendus*. In 1873, he published his retrospect of the (by then) forty years of labor: the book *L'architecture du monde des atomes*.

During all this time, Gaudin's conception of molecular geometry remained unchanged and fairly close to that which he had inherited from Ampère. Like Ampère, he interpreted the act of combination as being the formation of a polyhedral molecule; "the reason behind composition is a geometrical reason," as he put it.[40] These polyhedral molecules in turn built up crystals; hence crystal form was intimately related to molecular shape. But in detail, Gaudin's assumptions differed markedly from Ampère's.

With regard to molecular constitution, Gaudin dispensed with Ampère's assumption that all molecules were polyhedral for a simpler one in the case of elementary gaseous molecules. As for chemical combination, Gaudin abandoned Ampère's premise that molecular polyhedra simply interpenetrated each other while remaining more or less intact, in favor of his own theory that the atoms of constituent molecules were totally rearranged in combination to form a new, unitary, symmetrical molecule, in which no structural elements of the constituent molecules persisted.[41] Gaudin —along with Baudrimont, Laurent, and, eventually, Dumas—rejected the then-dominant electrochemical dualistic theories of chemical combination, which assumed the persistence of the electropositive and the electronegative constituents within the new compound molecule. Gaudin's program was also a much more thoroughgoing attempt at the amalgamation of chemistry and crystallography than had been Ampère's sketch. Despite his initial focus on the gas hypothesis, Gaudin was primarily interested in determining the molecular shape of crystallizable substances, through the use of data on their crystal form.

Starting with relatively simple ones, Gaudin with singleminded fervor worked out atomic models for a prodigious number of chemical substances over the years. The first model he said he had developed, and one of the simplest, gives an illustration of his procedure. It was his model for Fe_3O_4. Accepting the chemical formula, only one arrangement satisfied his symmetry criterion: an octahedron with square base, in which the three iron atoms were lined up symmetrically along a central axis and the oxygen atoms were placed at the four corners of a square plane of symmetry perpendicular to the axis.[42]

More complex molecules were built up in Gaudin's models of chains or rows of atoms, like the three-atom axial row in the Fe_3O_4 case, placed parallel to and symmetrically about a central axial row which was usually the longest. Within each row, the atoms were also strung out symmetrically on either side of a central atom—so that there could only be odd numbers of atoms in each row—and at equal distances from each other (fig. 2). In yet more complex molecules, the subsidiary rows in their turn served as sub-axes.

Gaudin was concerned to square his models of molecules with the results both of chemical and crystallographical analysis *via* his mediating principle of molecular symmetry. But it was this latter which was of overriding moment to Gaudin, usually to the subordination of chemical and even crystallographical considerations. Thus, he paid virtually no attention to the actual chemical properties of his molecules for clues about their atomic arrangement, and his principle of symmetry excluded molecules possessing the same symmetry as crystals in the rhombohedral, monoclinic, and triclinic systems, though he tried to work out structural models which would relate the molecular forms he had devised for substances which crystallized in these systems, with the crystal forms.[43]

While Gaudin's ideas by no means went unnoticed during his lifetime, they remained on the whole uninfluential. Partly, this was due to his being outside the scientific establishment in France. He never held a teaching post and he worked in isolation, although he received encouragement to persevere in his efforts from scientists like Dumas and the geologist Élie de Beaumont. His forty-year-long obsession betrays the outsider. Beyond his lack of connections, although perhaps not unconnected with it, lay his obstinacy in the face of scientific developments, particularly those in chemistry and crystallography. What might have been novel in the early 1830's was by the time of Van't Hoff and

---

[40] M. A. Gaudin, "Resumé général d'une théorie sur le groupement des atomes dans les molécules et les causes le plus intimes des formes cristallines," *Paris, Compte rendus* 45 (1857): p. 920.

[41] M. A. Gaudin, *L'architecture du monde des atomes*, pp. xiv–xv, 151–153. Gaudin saw his ideas as differing from Ampère's in that Ampère's were dualistic ("dualisme dans toute sa plenitude," p. xiv). In one of his rare references to personal contact with Ampère, he wrote in *L'architecture*:

"J'eus beau représenter à Ampère, et aussi à Berzelius beaucoup plus tard, lors de son passage à Paris, que le dualisme en chimie était une idée peu philosophique, ces deux hommes célèbres ne voulurent faire aucune concession sur ce chapitre." *Ibid.*, p. xv.

Gaudin also subsequently rejected Laurent's concept of the fundamental and derived radicals because he could not envision how an organic molecule could survive a substitution reaction with its structure intact, even though the atomic symmetry had been destroyed. At the same time, in *L'architecture*, he claimed that Laurent had asserted personally to him that molecules were composed of atoms "*agglomérés entre eux sans aucun ordre,*" like coins in a purse, *Ibid.*, p. 151. He commented:

"Je fus stupéfait d'une pareille manière de voir et n'ai pas été moins étonné d'apprendre qu'il avait publié depuis une théorie *des noyaux moléculaire*, qui n'est qu'un essai timide calqué sur mes principes énoncés de la façon le plus générale dès l'année 1832." *Ibid.*, pp. 151–152.

[42] *Ibid.*, p. 36.

[43] E.g. *ibid.*, pp. 154–167, 174–181.

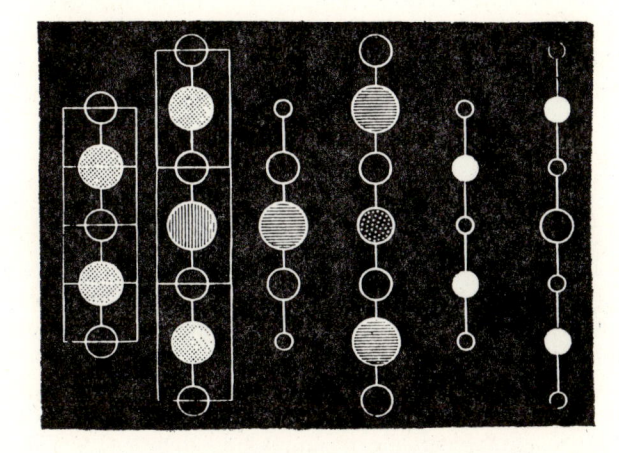

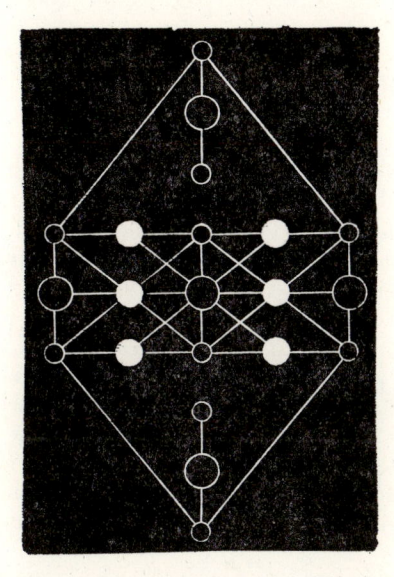

FIG. 2. Gaudin's models of atomic rows. The lower model is of a cane sugar molecule. *L'architecture du monde des atomes.*

LeBel an anachronism. Gaudin paid little or no attention to the intervening developments in organic chemistry, the concept of valence and the emerging stereochemistry of the 1860's.[44]

Even in the early 1830's, Gaudin's speculations were less obviously concerned with the major chemical issues of the day—with the exception, an important exception to be sure, of this interpretation of the gas hypothesis—than those of the other two chemists who oriented themselves toward the crystallographical-chemical concern with atomic arrangement, Baudrimont and Laurent. In particular, three developments of the previous decade dominated chemical thought and research in the 1830's: the elaboration of an electrochemical theory of chemical combination, the rise of organic chemistry, and the appearance of new structural principles, such as isomorphism, polymorphism, and the concept of isomerism, this last denoting substances which had the same apparent chemical composition but different physical and chemical properties, which suggested that atomic arrangement might be the key factor.

By the early 1830's, the dominant mode of explaining chemical phenomena was by means of one or another of the electrochemical theories, the most influential of which was the dualism of Berzelius. Arising out of the electrolytic experiments of the early decades of the century, these theories interpreted chemical attraction in terms of electrical forces (rather than the earlier gravitational or analogous forces). Particularly in Berzelius's dualistic theory, chemical combination was seen as the union of oppositely charged atoms or atomic groups to form a compound in which these constituents, although neutralized, somehow retained their identity. The paradigm was the formation of salts from electropositive bases and electronegative acids.[45]

By this time also, organic chemistry, swiftly developing its methods in the 1820's had reached a level of integration as a sub-science in its own right. Led by Berzelius, most chemists, whatever the complexities of their views regarding the relationship between organic and inorganic chemistry, nevertheless looked to the latter for guidance in the former, and this meant electrochemical dualism. Electrochemical dualism was certainly the context in which organic chemists thought about their subject in the early 1830's, and it was into

---

[44] In 1867 he was awarded the Prix Tremont by the Académie des Sciences, a prize for scientists pursuing worthy projects but lacking in funds. *Cf.* text of the award in *L'Architecture du monde des atomes*, pp. 229–231. His other achievements, such as his formula for silica and tin and titanium chloride, as well as his high temperature studies of aluminum and quartz and his preparation of artificial rubis, were also noted. For contemporary appraisals of his atomic speculations, *cf.* G. Delafosse, "Mémoire sur une relation importante que se manifeste, en certain cas, entre la composition atomique et la forme cristalline, et sur une nouvelle application du rôle que joue la silice dans les combinaisons minérales," *Paris, Mém. savans étrang.*13 (1852): pp. 546–548 (very unfavorable, *cf.* below for a discussion of this paper); A. Wurtz, *La théorie atomique* (2nd ed., Paris, Germer Baillière et Cie, 1879), pp. 186 and footnote 3, where Wurtz criticized Gaudin's inattention to chemical considerations; an anonymous review of his book in *Nature* 8 (1873): pp. 81–82, where he was criticized for not taking dynamical considerations into account. Van't Hoff, writing about attempts prior to his to establish the spatial arrangement of atoms, cited, as example, Gaudin's in *L'Architecture du monde des atomes* and commented: "Seulement il s'agissait de rendre accessibles à l'expérience les notions ainsi introduites." J. H. Van't Hoff, *Dix annés dans l'histoire d'une théorie* (Rotterdam, P. M. Bazendijk, 1887), p. 24, footnote 1. Le Bel actually invoked Gaudin's work in his paper, "Sur les conditions d'équilibre des composés du carbone," *Paris, Bull. Soc. Chim.* 3 (1890): p. 795.

[45] T. H. Levere, *op. cit.*, is the most recent and comprehensive treatment.

the dualistic model that they tried to fit their rapidly proliferating findings.[46]

In electrochemical schemes generally, and Berzelius's dualism in particular, the explanatory emphasis was on the electrochemical nature of the reactants. On the other hand, in the crystallographical-chemical approach, the explanatory key to chemical phenomena lay in structure and arrangement of the atoms. There was nothing necessarily contradictory between the electrochemical and the crystallographical-chemical approaches to explanation. Ampère himself offers the best example of their complementarity, since he not only was the formulator of a geometrical explanation of chemical combination but also, in the 1820's, an architect of electrochemical theory without, however, abandoning his earlier structuralist approach. Berzelius offers another example. Although the dominant electrochemical theorist, he had earlier tried to combine that theory with Haüy's crystallography-inspired mineralogy. He subsequently supported and encouraged Mitscherlich enthusiastically in the latter's work on isomorphism and polymorphism; he had made great use of the former principle in determining molecular formulas and hence, atomic weights. Finally, it was Berzelius who had defined, and even coined the term, "isomerism."[47]

However, in the hands of Baudrimont, first and most sweepingly, and then, in more complex fashion, Laurent, the crystallographical-chemical approach was raised in opposition to electrochemical dualism as an alternative mode of explanation of chemical union,[48] and the structuralist principles enunciated in the 1820's—particularly isomorphism—were employed by them in their elaboration of their alternatives. Moreover, although Baudrimont first formulated his alternative to electrochemical dualism mainly for inorganic chemistry, discoveries in the organic field, encapsulated in the so-called rules of substitution of Dumas, soon shifted the center of attention to organic chemistry, particularly in the work of Laurent. Here were to develop, during the remainder of the 1830's, the early structural ideas associated with Laurent's theory of fundamental and derived radicals and then the type theory.

Towards the end of 1833, Baudrimont published his first manifesto against electrochemical dualism and in favor of the crystallographical-chemical alternative: the book, *Introduction à l'étude de la chimie par la théorie atomique*.[49] Trained as a physician and a pharmacist, it is not known exactly how he found his way to an interest and training in crystallography. But however it happened, this preoccupation was evident in Baudrimont's first chemical publication, a paper of 1832 in the *Annales de chimie* in which he utilized a molecular model like Haüy's *molécules intégrantes* to explain certain characteristics of bismuth crystals as well as those of other substances.[50]

In the *Introduction* of the following year, Baudrimont delineated his theory of atomic arrangement, a subject which he, like Gaudin, was to pursue for almost another half-century. The purpose of his first book was to justify the theoretical basis for chemistry in atomism, even though he did not believe it was yet possible to express atomic formulas unambiguously. The work was divided into two sections: a theoretical first part, in which chemical phenomena were "deduced" from the atomic hypothesis and an inductive second part, in which the former procedure was reversed, the necessity for atomism being demonstrated by chemical phenomena.

The subsidiary themes which appear throughout the book were Baudrimont's orientation towards the crystallographical-chemical approach and his opposition to electrochemical dualism. With regard to the first of these, the work is saturated with crystallographical ideas, even to the use of the already somewhat archaic terms, *molécule intégrante* and *molécule constituante*.[51] Beyond Haüy, there is little doubt that Baudrimont's ideas also reflected those of Ampère and Gaudin. Later in his life, Baudrimont was to express his admiration for their molecular structural theories and acknowledge specifically the influence of Ampère's ideas on his own.[52]

To Baudrimont, the *molécule intégrante* was formed by the symmetrical arrangement of its atoms and hence was of regular and determinate form. These built up

---

[46] *Cf.* J. H. Brooke, "Organic Synthesis and the Unification of Chemistry—A Reappraisal," *Brit. Jour. Hist. Sci.* 5 (1970–1971): pp. 378 ff.

[47] *Cf.* below, chap. 5.

[48] S. C. Kapoor, "The Origins of Laurent's Organic Classification," *Isis* 60 (1969): pp. 477–527, especially 491 ff. *Cf.* also T. H. Levere, *op. cit.*, pp. 169 ff. and J. H. Brooke, *Brit. Jour. Hist. Sci.* 5 (1970–1971): pp. 378 ff. and N. W. Fisher, "Organic Classification before Kekulé," *Ambix* 20 (1973): pp. 112 ff. for discussion of this issue.

[49] (Paris, Louis Colas, 1833).

[50] A. E. Baudrimont, "Recherches sur la forme des atomes," *Annal. de chimie* 50 (1832): pp. 198–205. In the dedication of the *Introduction* (to Thenard), Baudrimont indicated that he had been concerned with atomic speculations for some time (no pages indicated).

[51] A. E. Baudrimont, *Introduction*, pp. 22–23, 82–83. *Cf.* also pp. 9, 85.

[52] *Ibid.*, p. 75, for gas hypothesis which he wrote served as "de base à M. Gaudin pour déterminer les poids de plusieurs atomes." There is brief reference to Gaudin's molecular models on p. 145. For subsequent references to the influence of Ampère's ideas on his own, *cf.* "Théorie des substitutions.... Réclamation de M. Baudrimont," *Revue scientifique* 1 (1840): pp. 35–36; *Traité de chimie générale et expérimentale* (2 v., Paris, J. B. Ballière, 1844–1846) 1: pp. 9, 115. For high praise of Gaudin's speculations, *cf. Thèse sur cette question: Quel est l'état actuel de la chimie organique et quels secours a-t-elle reçus des recherches microscopiques?* [Faculté de Médecine de Paris, Concours pour une chair de pharmacie et de chimie... soutenue le 20 mars 1838] (Paris, Paul Renouard, 1838), p. 108. *Cf.* also the last of his many memoirs on molecular structure for praise of Ampere and Gauden: "Cinquième mémoire sur la structure des corps," *Bordeaux, Mém. Soc. sci. phys.* 2 (1878): pp. 140–149.

crystals also of specific and regular shapes. However, although there existed an immense number of chemical combinations, the number of crystalline primitive forms was much more limited. In order to account for similarity of primitive forms possessed by crystals of different composition, Baudrimont invoked Mitscherlich's principle of isomorphism.[53]

So far, Baudrimont was quite conventional, but he introduced some novelties of his own into this principle by employing isomorphism in a greatly extended manner compared to its usual use. Thus, although Mitscherlich had originally defined isomorphism solely in terms of atomic arrangement, he had come to apply it to series involving the substitution of chemically similar isomorphic components. Baudrimont, however, reverted to the original definition. He suggested that any compounds, whatever their chemical nature, could be grouped in the same isomorphic series provided only that they crystallized in the same form and, presumably, their individual formulas were subsumable under a general representative formula. As he put it:

One could add: 1° That Cu^2O cristallizes as I^2K and the other isomorphic combinations of iodine; that, consequently copper ought to be isomorphic with iodine, and potassium with oxygen; 2° That S Ba cristallizes as Cr K; that, in this case, chromium is isomorphic with sulfur, and the oxide of potassium with that of barium; ... One could say the same for all the substances of the same series; indeed, the formula $\overset{..}{C}a \vartriangle$ already used, can be transformed into $\vartriangle \overset{.}{\vartriangle}$ and those substances so composed conserve the same form.[54]

The striking implication in this generalization was that the arrangement of atoms, as revealed by crystal form, rather than their chemical nature, was the key to an understanding of chemical phenomena and a classification of chemical substances.

Another major theme of the book, related to the first, was Baudrimont's opposition to electrochemical dualism. Baudrimont criticized this theory in terms of what it implied about atomic arrangement within the chemical molecule; namely, that there persisted two constituent components—either individual atoms or atomic groups—of opposite electrical charge. To him, this implication was unprovable and, moreover, erroneous. In his arguments against it, Baudrimont appealed to electrolytic evidence itself for support: that electrolytic decomposition did not necessarily decompose a compound substance into its dualistic components (which, by implication, must therefore have preexisted in the compound molecule), but rather, a variety of products of decomposition could be obtained, depending upon the intensity of the pile used.[55]

But, beyond this, Baudrimont's rejection of dualism centered upon his belief that chemical reactivity involved atomic movement in which any arrangement of atoms in the constituent molecules would be destroyed in the process of combination. Thus, no study of chemical reactions *per se* and, particularly, no electrolytic evidence could provide indication of atomic arrangement. The only possible way of obtaining such information was through study of crystal forms: "It is only through *geometrical observations made on crystals and through mathematical considerations* that this goal [the elucidation of atomic arrangement] can be attained."[56]

Baudrimont provided a fair number of examples of the sort of thing he had in mind in the *Introduction*. The context in which they were introduced was rendered somewhat ambiguous by a third major theme of the work: the futility of trying to establish true molecular formulas from gravimetric data. Indeed, at the very end of the book, Baudrimont concluded that chemists ought to abandon molecular formulas in favor of chemical equivalents, precisely because the former were unknowable.[57]

But in order to demonstrate his negative proposition about molecular formulas, Baudrimont turned to speculation about possible atomic arrangements: both common salt (Cl^2Na in his formula) and galena (S Pb) crystallized as cubes and therefore their *molécules intégrantes* had to be cubic. However, from these *simplest* percentage formulas it was impossible to envision the construction of cubic molecules. But if one cubed the numbers of atoms in these formulas, the situation was clarified. Eight atoms of Pb S and twenty-seven of Cl^2Na could be arranged to form cubic molecular structures.[58] Some substances, like camphor, both had formulas containing a perfect cubical number of atoms ($C^{10}H^{16}O$) and crystallized in the cubic form. Baudrimont seized on this case to work out a cubic *molécule intégrante:* the oxygen atom in the center, six carbon atoms at the centers of each face, the other four at the apices of four of the solid angles, twelve of the hydrogen atoms occupying the centers of the twelve edges and the remaining four marking the apices of the four still-unoccupied solid angles.[59]

This last case points up Baudrimont's ambiguity, for if these models had been pressed into service to

---

[53] A. E. Baudrimont, *Introduction*, pp. 26–27.
[54] *Ibid.*, pp. 28–29.
[55] *Ibid.*, pp. 41–42.
[56] *Ibid.*, pp. 50–51.
[57] *Ibid.*, p. 202.
[58] For Cl^2 Na:

Na Cl Na   Cl Cl Cl
Cl Cl Cl   for the outer layers Cl Na Cl   for the intermediate layer
Na Cl Na   Cl Cl Cl

Ibid., p. 31 for the PbS model as well.

[59] *Ibid.*, pp. 32–34. *Cf.* also S. C. Kapoor, *op. cit.*, p. 499 for the Cl^2Na model (which he gets slightly incorrect, calling the outer layers "first and third faces" and the intermediate one "major face") and his penetrating observation on the similarity between Baudrimont's model of camphor and the organic models later devised by Laurent.

demonstrate the futility of deducing molecular formulas from the results of chemical analysis, nevertheless Baudrimont concluded about camphor that:

> The chemical formula of camphor exactly expressed the number of atoms entering into its *molécule intégrante,* and the result is that the preceding observations [the speculation about the atomic arrangement in the camphor molecule] give a mathematical confirmation to the analysis of this substance made by Dumas.[60]

It seems clear, from what Baudrimont wrote in the *Introduction* as well as from his subsequent interests, that he was in fact setting forth a crystallographical-chemical approach as the means—indeed, as the only means—of ascertaining true atomic arrangement within the chemical molecule. Baudrimont also gave some consideration to organic chemistry in the *Introduction*. Since he saw no disjunction between the organic and the inorganic realms, he saw no reason why organic chemistry should not also be amenable to the same kind of structural treatment which he envisioned for inorganic chemistry:

> 1° The only difference which can be established between inorganic and organic chemistry relates only to the origin of the substances considered by these sciences. 2° The cristallizable substances, of organic origin, can be compared with those of inorganic origin in all respects; they are composed of atoms arranged among themselves to form bodies of regular form and definite proportions and often indeed these atoms are of similar natures. . . .[61]

Until recently, at least, Baudrimont has been all but forgotten by historians of science. During his own lifetime—a life not without moderate success and happiness—his devotion to speculating about atomic arrangement went largely unnoticed. His own eulogistic biographer, writing almost immediately after his death, did not see fit to discuss these speculations.[62]

Yet his anti-dualist, structural approach to chemical theory actually appeared at a propitious time in the development of organic chemistry. For in 1834, the year after Baudrimont's *Introduction* appeared, Dumas published the summary of his discovery that constituents of organic compounds could be replaced by others of very different chemical nature, codified in his three rules of substitution:

> 1°. When a substance containing hydrogen is submitted to the dehydrogenating action of chlorine, of bromine, of iodine, of oxygen, etc., for each atom of hydrogen which it loses, it gains an atom of chlorine, or bromine, of iodine, or half an atom of oxygen.
> 2°. When a compound contains oxygen, the same rule holds good without modification.
> 3°. When the hydrogenized body contains water, this loses its hydrogen without replacement and then, if a further quantity of hydrogen is removed, it is replaced as in the former case.[63]

Dumas himself claimed nothing more than the status of "empirical laws" for his substitution rules, but they did have quite striking theoretical implications for a fundamental tenet of electrochemical dualism: that the electrochemical nature of a reactant governed its role in a chemical reaction and its function and position in the resultant chemical compound. In Dumas's substitution rules, the electrochemical nature of the reactants seemed to be (or could be interpreted as being) relatively unimportant. Strongly electronegative elements like the halogens could replace strongly electropositive ones like hydrogen.

Yet Dumas himself was by no means an anti-dualist in 1834. Indeed, he was governed in the researches which resulted in the rules of substitution by dualistic models, especially that of alcohol as a binary compound of a hydrocarbon and water.[64] And in the very paper in which he set forth his substitution rules, he roundly criticized two unnamed "speculative" chemists, one of whom was certainly Baudrimont, for espousing anti-dualistic notions, characterizing their views as follows:

> It [their position] rests on ideas relating to the form and the arrangement of the molecules of substances, and in the opinion of these two authors, it would be impossible to represent the mechanical structure of compound atoms by admitting the persistence of binary compounds, as when an acid and a base form a salt, for example.[65]

To this, Dumas replied:

> I shall thus admit, until proof to the contrary, that the elements of binary substances conserve their disposition in saline combinations. I add that the system of ideas which I have adopted touching on organic nature, and which consist in assimilating its combinations to those of mineral chemistry, remains a virtual stranger to these discussions [of the utility of the speculations of these anti-dualistic chemists], so long as they put forward [the theory] that predisposed molecular arrangements are not admissible in either the one or the other of these two classes of substances.[66]

If Dumas remained committed to electrochemical dualism after enunciating the substitution rules, others pursued their implications. One of these was Baudrimont himself. Although he was not actively engaged in organic research during the 1830's, an opportunity

---

[60] A. E. Baudrimont, *Introduction*, pp. 33–34.
[61] *Ibid.*, pp. 63–64.
[62] L. Micé, "Discours d'ouverture de la séance publique du 19 mai 1886 (éloge de M. Baudrimont)," *Bordeaux, Actes Acad. Sci.* **42** (1880): pp. 729–766; "Éloge de M. Baudrimont, notes complémentaires," *ibid.* **44** (1882): pp. 557–624.

[63] J. B. Dumas, "Considérations générales sur la composition théorique des matières organiques," *Jour. de Pharm.* **20** (1834): pp. 284–285. Translation from J. R. Partington, *A History of Chemistry* (4 v., London, Macmillan, 1961–1964) **4**: p. 361.
[64] *Cf.* S. C. Kapoor, *Ambix* **16** (1969): pp. 52 ff. and *Dictionary of Scientific Biography* **4**: p. 245 for background to the laws of substitution.
[65] J. B. Dumas, *Jour. de Pharm.* **20** (1834): p. 264. *Cf.* also pp. 262–263.
[66] *Ibid.*, p. 266.

for surveying developments in that subject was provided by a competition in which he took part for the chair of pharmacy and organic chemistry at the Faculté de Medicine in 1838, for which Baudrimont had to submit and defend a thesis.[67]

Baudrimont's thesis was entitled, *Quel est l'état actuel de la chimie organique et quels secours a-t-elle reçus des recherches microscopiques?* The latter subject received short shrift; it was organic chemical theory in which Baudrimont was interested, and specifically, Dumas's rules of substitution. Given the set of ideas expressed in the *Introduction* of 1833, it is not surprising that Baudrimont should have considered the substitution rules to be merely an extension—a verification almost—of his own earlier ideas, particularly his generalization of the principle of isomorphism to include substances of different chemical natures.[68] Generalizing broadly from this and from his belief that the chemical molecule was a unitary, not a dualistic entity, Baudrimont suggested a classificatory method for organic substances which would be based on the similarity of what he termed the "general formula."[69] But by 1838, far more comprehensive and significant work in this direction had already been carried out by a chemist who had been Dumas's pupil and assistant at the time Dumas had formulated his rules of substitution: Auguste Laurent. Using substitution as his basis, Laurent developed the implication of Dumas' rules in the direction of structural organic chemistry by means of the crystallographical-chemical approach we have been tracing. Indeed, Laurent's work was perhaps the most sophisticated and fruitful exemplar of this tradition.

It is not difficult to discover where Laurent learned his crystallography. Born in 1807, he received his professional education at the École des Mines, entering in 1826 and receiving his diploma in 1830.[70] Haüy himself had taught here, and the tradition of Haüy's crystallography was strong. Crystallography and crystallographical models played a great role in Laurent's work. Laurent taught crystallography and even published a short textbook on it late in his career. Alexander Williamson, in his obituary of Laurent, pointed out the structuralist implication of Laurent's crystallographical orientation for his work: "At an early period also, he had laid down the fundamental proposition, that 'Form and arrangement may be as important as composition,' a proposition round which all his researches may be said to gravitate."[71]

The actual chemical investigations which led Laurent to the elucidation of atomic arrangement were his studies of the reactions of naphthalene with the halogens and nitric acid. After receiving his diploma from the École des mines, he accepted a position as assistant to Dumas who introduced him to this subject in 1831. He soon produced a series of chlorinated, brominated, and nitrated naphthalene products. In this period, Dumas himself was moving towards the enunciation of his substitution rules and these played an important guiding role in Laurent's early work on naphthalene compounds although later, in the heat of polemic against Dumas, Laurent was to deny this.[72] But within a year or two after the appearance of Dumas's substitution rules—in 1835 and 1836—Laurent had passed beyond Dumas towards a structuralist approach to organic chemistry. The key to his advance lay in his formulation of the concepts of "fundamental and derived radicals," by which he incorporated the idea of substitutions embodied by Dumas's rules into a general structuralist model applicable to the whole of organic chemistry.

The concept of the "radical"—an atomic group which persists unchanged through a series of chemical reactions and which therefore functions as a chemical element—dates back to the time of Lavoisier. In the second and third decades of the nineteenth century, evidence had accumulated for the existence of simple radicals, like the cyanogen radical discovered by Gay-Lussac in 1815 (CN). By the mid-1830's organic chemists, led by Dumas and the German chemists, Justus von Liebig and Friedrich Wöhler, had made the radical concept the focus of organic chemical theory and had used it as the means of extending electrochemical dualism to organic chemistry, since organic compounds could be formulated dualistically with the radical as one of the two units.[73]

---

[67] His main (and successful) competitor for the chair was Dumas.

[68] A. E. Baudrimont, *Quel est l'état actuel de la chimie organique* . . . , p. 104.

[69] *Ibid.*, p. 117. Baudrimont went on to suggest such formulas for large numbers of inorganic compounds (eg., II X $^4\Delta$ for sulfates, seleniates, chromates, molybdates tungstates etc.) and the formula II^{4}X^6 for ethers and substitution products (including aldehyde and chloral). He later cited this passage, specifically, in his reclamation against Dumas over the type theory. *Cf.* A. E. Baudrimont, *Revue scientifique* 1 (1840): pp. 37–43, especially p. 43.

[70] S. C. Kapoor, *Isis* 60 (1969): p. 477; "Laurent, Auguste (or Augustin)," *Dictionary of Scientific Biography* 8: pp. 54–61.

[71] "President's Address," *Chem. Soc. Jour.* 7 (1854): p. 155. For ascription to Williamson, *cf.* C. de Milt, "Auguste Laurent—Guide and Inspiration of Gerhardt," *Jour. Chem. Ed.* 28 (1951): p. 198.

[72] *Cf.* the paper in which Laurent first enunciated his concepts of fundamental and derived radicals: "Sur le nitronaphtalase, la nitronaphtalèse et la naphtalase," *Annal. de chimie* 59 (1835): pp. 388 ff. For his denial of the influence of Dumas's substitution laws, *cf.* his "Réclamations de M. Laurent," *Revue scientifique* 1 (1840): pp. 53–54. A more balanced appraisal by Laurent is in his "Lettre de M. Laurent sur les substitutions," *ibid.* 2 (1840): p. 55.

[73] *Cf.* above, footnote 46 and A. J. Ihde, *The Development of Modern Chemistry* (New York, Harper and Row, 1964), pp. 184–191 for general background.

Characteristically, Baudrimont, with his strictures against the persistence of constituent atomic complexes in compound molecules and against electrochemical dualism generally, had rejected the organic radical in his *Introduction*.[74] Laurent, on the other hand, adapted the radical concept to his own structuralist uses. By 1836, he had come to see the organic chemical molecule as structured into two components: an inner hydrocarbon, the fundamental unit of the compound—its radical—to which it owed its basic chemical nature. But Laurent combined this with the substitution concept. As he defined his "fundamental" and "derived radicals" in 1836:

> All organic combinations are derived from a hydrocarbon, the *fundamental radical,* which often [itself] no longer exists in these combinations, but is represented there by a *derived radical,* which contains as many equivalents as it [the fundamental radical].[75]

Thus, the inner hydrocarbon of fixed number of carbon atoms, the fundamental radical, could itself be transformed into derived radicals in substitution reactions, whereby the associated hydrogen atoms were replaced by atoms of other elements, such as the halogens or oxygen.

However, the inner radical could unite with atomic or molecular units which remained outside of it, these latter produced by addition reactions, rather than substitution reactions, and less intimately bound in the compound.[76] Laurent's model might seem at first glance little different from orthodox electrochemical dualistic interpretations. But his use of the substitution concept indicated the essential difference between his and the electrochemical dualists: Laurent emphasized formal, structural orientation over the electrochemical nature of the constituents or the forces which effected chemical combination. This was made clear, for example, when Laurent discussed the issue of the properties which addition components (added outside the radical) gave to an organic compound:

> If a dehydrogenizing substance, such as oxygen, shall enter into combination, but outside the radical, it will render it acid, however great be the quantity of carbon, hydrogen and oxygen making up the radical, and however small the quantity of oxygen outside the radical be; the acidity depending not at all on the rapport of carbon and hydrogen to the oxygen, but only on the place of this latter.[77]

Thus, Laurent's organic molecular model of the mid-1830's was a kind of fusion or synthesis of electrochemical dualism with the approach which focused upon atomic arrangement as the key to understanding chemical phenomena. Laurent himself was perfectly clear about this, as he showed in his first comprehensive survey of his ideas on molecular structure and his earliest surviving speculations on atomic arrangement. These were found in the *Introduction* to the first part of the doctoral dissertation which he defended in 1837. This introduction, entitled "Théories des combinaisons organiques," was built around a consideration and critique of what Laurent distinguished as the two approaches to the elucidation of organic chemistry: electrochemical dualism and the crystallographical-chemical approach, the latter being characterized as follows:

> In the second system, one explains the formation of compound substances by supposing that several atoms of one substance unite with several atoms of another to form a new molecule in which the atoms are symmetrically grouped according to a certain regular geometrical form without any atom being combined more [intimately] with one or another atom than with all the rest. Such an assemblage can only be represented by *une formule brute*.[78]

How familiar Laurent was with the whole tradition of this sort of speculation going back to Ampère is problematical. That he could have been aware of Ampère's earlier speculations is most likely, especially since he had worked with Dumas. And, in addition, there is the fact that Laurent came to espouse the gas hypothesis subsequently, associating both it and the divisibility of atoms, when he did, with Ampère.[79] But in the 1837 thesis, there was no mention of Ampère (nor of Gaudin).

Rather, it was Baudrimont who figured here, and in very prominent fashion, as the exemplar of the tradition of crystallographical-chemical geometry. That Baudrimont served as an intermediary in transmitting this tradition to Laurent was subsequently attested to by Laurent. He wrote, in a note of reclamation against Dumas's type theory, published in 1841, that he had been led to structural models of atomic arrangement by his desire to "harmonize the ideas of M. Baudrimont with those in vogue."[80]

---

[74] A. E. Baudrimont, *Introduction*, p. 80.

[75] A. Laurent, "Théorie des combinaisons organiques," *Annal. de chimie* **61** (1836): p. 126 (Deuxième proposition).

[76] The central radical (fundamental and derived) always contained the same number of chemical equivalents. The addition products, outside the radical in Laurent's model, added to the number of equivalents of the compound and was relatively easily removed, e.g. by heating with caustic potash. *Cf.* S. C. Kapoor, *Isis* **60** (1969): pp. 507 ff., especially for Kapoor's observation on Laurent's "error" in developing this idea for naphthalene and other aromatics, where the model did not in fact apply.

[77] *Ibid.*, pp. 127–128. In this same paper, Laurent proposed the existence of chloroacetic acid.

[78] This introduction, not appended to the original published text of the thesis, was published by J. Jacques, "La Thèse de Doctoral d'Auguste Laurent et la théorie des combinaisons organiques (1836)," *Bull. Soc. Chim. France* (1954): D. 34–39. The quotation is from D. 35.

[79] His first treatment of atomic divisibility was in the paper, "Série naphthalique. 31 mémoire sur les types," *Revue scientifique* **14** (1843): pp. 96–104, but his concern here was not with the gas hypothesis as such, but with considerations of chemical structural formulas and related matters which we shall take up presently. Also, there was no mention of Ampère in this paper.

[80] "Lettre de M. Laurent," *Revue scientifique* **5** (1841): p. 123. *Cf.* S. C. Kapoor, *Isis* **60** (1969): p. 492, footnotes 50, 51.

This summarized concisely the course of Laurent's argument in the introduction to the first part of his thesis. Throughout his career, Laurent's feelings towards Baudrimont were ambivalent. The two men were almost the same age and in the 1830's and early 1840's, both had in common the frustrated wish to obtain secure teaching positions in Paris. Coincidently, Baudrimont succeeded Laurent in his provincial post at the University of Bordeaux in 1847, after the latter had abandoned the position for a last try in Paris. In print, Laurent tended to defend Baudrimont's originality, especially if it were at the expense of Dumas. Privately, Laurent could write about Baudrimont: "Baudrimont is a dreamer who doesn't lack for ideas, but who takes no account of experiment and is not at all interested in developments in organic chemistry."[81]

This ambivalence characterized Laurent's treatment of Baudrimont in his thesis. While Baudrimont was the only chemist cited by Laurent to illustrate the crystallographical-chemical tradition (and Laurent set forth in detail the models for sodium chloride and lead sulfide which Baudrimont had given in his *Introduction*) Baudrimont's models were not given by Laurent as emulative examples, but rather to be criticized for their arbitrariness and ambiguity. He wrote bitingly of the crystallographical-chemical tradition, by way of introduction to Baudrimont's molecular models:

This one, to compensate for its sterility, pretends to have the advantage over the other theory [electrochemical dualism] of being able to reconcile the crystalline form of compounds with those of [their] constituents: a pretention without basis, for the [facts] on which it rests do not bear up against examination.[82]

Laurent also rebutted Baudrimont's argument that the persistence of constituent atomic groups in the compound molecule was destroyed by the atomic movements of the reaction itself—an argument which Laurent admitted was a strong one—by citing the persistence of a characteristic color of a constituent group in a series of compounds.[83]

Laurent's own program was "to reconcile the two preceding [approaches] in conserving the advantages which each of them presents." In connection with this synthetical program, meant both for inorganic as well as organic chemistry, Laurent elaborated upon his fundamental and derived radical concept by means of a crystallographical model:

I shall imagine a given hydrocarbon, $C^{12}H^{12}$, if you wish, and I shall represent it by a prism of six sides, having at the twelve solid angles, twelve atoms of carbon, and in the middle of the edges of its bases, twelve atoms of hydrogen. This prism could be surmounted by two pyramids as modifications, one on each base. If the pyramid is of water, the formula will be $C^{12}H^{12} + 2HO^{\frac{1}{2}}$ or $2\ H^2O$, and it will represent an ether or an alcohol. If these pyramids are replaced by others of sulfuric acid, hydrochloric acid etc., . . . salts will be formed which ought to be represented by an hexahedral prism plus the modifications of the bases, that is, by [the formulas] $C^{12}H^{12} + SO^3$, $H^2\ Cl^2$, $Az^2\ O^5$, etc. . . .

One could transform an hydrated salt either by placing on the bases, the *modifications* composed of water and acid, or by leaving the acid on the bases and disposing the molecules of water symmetrically on the faces of the prism; in the first case the formula will be $C^{12}H^{12} + 2(SO^3 + H^2O)$ and in the second $2,3,4,6\ HO^{\frac{1}{2}} + C^{12}H^{12} + 2SO^3$.

Thus one sees that, just as in crystallography, one could remove the modifications, or a part thereof, without replacing them, or replacing them wholly or partially.

But, it will not be the same in the central frame, or in the radical; one cannot remove from it a sole piece—that is to say—a sole atom without destroying it, at least if one does not replace the removed atom by another equivalent atom which will continue to maintain the equilibrium of the frame.[84]

Laurent's model, apart from its insistent retention of a dualistic scheme, differed in its theoretical status from the kinds of models of atomic arrangement which Gaudin and even Baudrimont had devised by its analogical nature. Laurent was not inferring from the specific crystal form of any individual substance the form of that substance's molecules; rather, in taking crystal structure generally as an analogy, Laurent was suggesting a possible manner of envisioning organic molecules (again in a generalized sense), and their transmutations in chemical reactions.[85]

---

[81] "Lettre 10: Laurent à Gerhardt, Bordeaux, 11 mai 1845," *Correspondance de Charles Gerhardt,* ed. M. Tiffeneau, tom I: *Laurent et Gerhardt* (Paris, Masson et Cie, 1918), p. 44. For Laurent's defense of Baudrimont's priority against Dumas, *cf.* "Lettre de M. Laurent," *Revue scientifique* 1 (1840): p. 343.

Laurent, in retrospect, credited Baudrimont with two innovations: his anti-dualism and his organizing inorganic salts into types. *Cf. Méthode de chimie* (Paris, Mallet-Bachelier, 1854), p. 358. Laurent felt, however, that Baudrimont had carried his opposition to atomic predisposition in molecules too far. *Cf.* also "Classification chimique," *Paris, Comptes rendus,* 19 (1844): p. 1099.

[82] J. Jacques, *Bull. Soc. Chim. France* (1954): D. 35. To electrochemical dualism, Laurent put his objection in crystallographical terms: he could not see how two molecules, each of a specific form, could give rise to a third, with a form incompatible with those of its constituents, and with these latter continuing to exist in the compound molecule. This was similar to Baudrimont's argument in the *Introduction,* p. 80.

[83] J. Jacques, *Bull. Soc. Chim. France* (1954): D. 38.

[84] *Ibid.,* D. 39. Earlier he wrote:

"Dans la théorie que je vais exposer, j'ai cherché a consilier les deux précédentes, en conservent les avantages que chacune d'elle présente." *Ibid.,* D. 35. In the body of the published dissertation, Laurent expressed similar sentiments but was much more sympathetic to Baudrimont's views. A. Laurent, *Recherches diverses de chimie organique. Sur la densité des argiles cuites a diverses temperatures* [Thèse de chimie et de physique présentée à la faculté des sciences de Paris le 20 décembre 1837] (Paris, E. J. Bailly, 1837), pp. 113–114.

[85] Thus, about a similar complicated (and not really feasible) model for lead sulfate, Laurent commented that he was "loin de prétendre que les atomes soient arrangés ainsi dans les sulfates de barytes etc." J. Jacques, *Bull. Soc. Chim. France* (1954): D. 35. *Cf.* also, A. Laurent, *Paris, Comptes rendus* 19 (1844): pp. 1098–1099.

Yet, to say that crystal structure merely served as an analogy for Laurent would also place too narrow an interpretation on his creative scientific thought in this period. If he never went as far as Gaudin in inferring specific molecular form from crystal structure, he certainly made the attempt, in much of his ensuing work, to combine crystallographical and chemical data in the interest of developing his structuralist approach.[86] The guiding concept in his research here was to be isomorphism, organic substitution reactions being viewed as analogous to inorganic isomorphic replacement. In a series of organic compounds which Laurent regarded, on chemical grounds, as being derived from the same fundamental radical, the molecular structure common to them all ought to persist and be discernible in the crystal structure of such series that produced crystallizable compounds. Laurent had already postulated such isomorphic relations in his 1837 thesis.[87]

Laurent was not alone in recognizing the implications for isomorphism in the rules of substitution. Baudrimont, after all, had used the relation of substitution to isomorphism in his thesis of 1838 in an effort to belittle the originality of Dumas's rules and Dumas himself, in the following year, 1839, abandoning his commitment to electrochemical dualism, and incorporating his own discoveries of substitution reactions into the very heart of his chemical theory, arrived at a structuralist and unitary concept of the organic molecule quite similar to Laurent's (and Baudrimont's). This was embodied in his concept of organic "types," chemical and mechanical. The former contained "the same number of equivalents united in the same manner" and was "endowed with the same fundamental chemical properties"; the latter was closer to what Baudrimont had defined as the "general formula" in his thesis of 1838. We shall not deal with Dumas's theory in any detail, important though it was for the theory of organic chemistry, since the conceptual context of Dumas's thought at this time lay largely outside the crystallographical-chemical approach. But it must be pointed out that in his formulation of the concept of "type" Dumas realized that his own rules of substitution in organic reactions were analogous to the principle of isomorphism in inorganic chemistry. Thus, he wrote in his first paper on chemical types, cautiously as was his wont:

... in organic chemistry, the theory of substitution plays the same role as isomorphism in mineral chemistry and, perhaps some day, it will be found by experiment that these two general views are closely linked, derive from the same case, and can be generalized under a common expression.[88]

But it was Laurent who pushed forward the study of the isomorphism of substitution products through the comparison of their crystal forms. Laurent was forestalled in the first measured determination of such isomorphism by a paper of the crystallographer, Frederic Hervé de la Provostaye, in 1840.[89] But Laurent had postulated isomorphic relations in his thesis of 1837, and in 1839 he had submitted a paper to the Académie des Sciences on the isomorphism of various chlorine and bromine derivatives of naphthalene, which, however, remained unpublished.[90] Laurent's actual publications on the subject began with a note of 1841, in which he asserted that: "... when hydrogen is replaced by its equivalent of chlorine, bromine, oxygen, nitrous acid, etc., the *formula* and the *form* of the new compound are similar to those of the substance which gave birth to it.[91]

Between then and 1845, Laurent extended and refined his conception of the isomorphism of substitution by the continued study of the crystal forms of substitution products. Thus, already in 1841, he had written that the correlation between formula and form was not necessarily applicable to all cases; it could be modified by the occurrence of dimorphism and isomerism, for example, as well as by something which, Laurent wrote cryptically, "I [shall] indicate it in my memoir." What he apparently had in mind was taken up in a publica-

---

[86] In the physics thesis submitted along with the one in chemistry in 1837 and titled, "Considérations générales sur les propriétés physiques des atomes et sur leur formes," Laurent had outlined a comprehensive research program for ascertaining crystal structure and ultimately atomic arrangement through crystallographical, chemical and physical (optical and electrical) data. A. Laurent, *Recherches diverses de chimie organique*, pp. 118–119.

[87] A. Laurent, *Recherches diverses de chimie organique*, p. 11.

[88] J. B. Dumas, "Mémoire sur la constitution de quelques corps organiques, et sur la théorie des substitutions," *Paris, Comptes rendus* 8 (1839): p. 621.

[89] J. F. Hervé de la Provostaye, "Isomorphisme de l'oxaméthane et du chloroxaméthane," *Annal. de chimie* 75 (1840): pp. 322–325.

Cf. Mitscherlich's extremely interesting comments on De la Provostaye's work: he was not completely convinced that De la Provostaye had proved his point, although he guardedly accepted the possibility of isomorphism between substitution products. However, he also emphasized that the nature as well as the place of the atoms was important. E. Mitscherlich, "Ueber die chemische Verwandschaftskraft," *Gesammelte Schriften*, pp. 489–491.

[90] But summarized in a note of reclamation against De la Provostaye, "Sur l'isomorphisme de certains corps liés entre eux par la loi des substitutions," *Paris, Comptes rendus* 11 (1840): p. 876. In his "Troisième mémoire sur la série du phènyle (extrait)," Laurent wrote: "... je avais remarqué plusieurs anomalies, et j'ai retardé l'impression de ce Mémoire, parce que j'espère pouvoir bientôt les expliquer. En tout cas, je n'ai pas donné des resultats numériques de mes mesures," *Revue scientifique* 9 (1842): p. 23.

[91] A. Laurent, "Sur les acides nitrobromophénisique et ampélique, etc., *Paris, Comptes rendus* 12 (1841): p. 612. He had demonstrated isomorphism between a large number of benzene derivatives (phenyl series) and had concluded that the substitution constituents must also be isomorphic with each other: chlorine with hydrogen, nitrous acid and bromine, KO and "oxide of ammonium" with water (or K and ammonium with H). *Cf.* also below, chap. 6, for De la Provostaye's studies of this question.

tion of the following year, 1842, and christened with the awkward neologism, "hemisomorphism." The specific case was the following: two naphthalene derivatives crystallized in different systems of symmetry even though they appeared to be isomorphic from the formulas as Laurent deduced them from their chemical reactions (granting isomorphism between chlorine and hydrogen). One crystallized as an oblique prism with rhombic bases (the triclinic system); the other as a right prism with rhombic base (the monoclinic system). The issue was how to reconcile the incompatible crystal forms with the apparent isomorphism of the organic molecules.

In order to resolve it, Laurent suggested a novel approach, one which demonstrated most clearly his manner of reasoning from crystallography to chemical molecular structure and back to crystallography. He used a model similar to the one he had given in his 1837 thesis to illustrate his radical concepts. This time, however, the model was meant as more than an analogical mode of visualizing his radical theory, if still less than an actual model of the physical molecule. It actually served as a guide to crystallographical measurement and interpretation:

From the chemical point of view, I think that the barrier raised between the different crystalline systems ought to be removed. Let us suppose that, in a hydrocarbon, the atoms are grouped cubically, and that each face carries an atom of hydrogen. If two opposite atoms of hydrogen are replaced by two atoms of chlorine, these latter can elevate or depress slightly the cube in one direction, and one will obtain a prism with a square base, only slightly different from the cube. If four atoms of hydrogen are replaced by four atoms of chlorine, one will obtain a prism with rectangular base little different from the prism with the square base or from the cube. Finally, if one replaces the six atoms of hydrogen with six atoms of chlorine, the original symmetry will be re-established, and one will have a new cube, a little bigger or a little smaller than the first; but the general disposition of the atoms will be the same in the first cube, the prism with square base, the rectangular prism and the second cube. One could say that these four compounds are isomorphs, if there be little difference between the lengths of their axes.[92]

After suggesting that this was the case with his two naphthalene derivatives on the basis of goniometric measurements, Laurent concluded: "However, one cannot say that these two substances are absolutely isomorphic. They have several things in common; so I shall propose the name of hemisomorphism to designate the demisimilitude of form which exists between the two compounds."[93]

By 1845, Laurent had overcome his scruples against denoting such cases as isomorphic. He had published his new system of organic classification in which compounds were arranged according to their hydrocarbon *noyau* or nucleus—the term which Laurent had come to substitute for fundamental radical—and which emphasized most explicitly that his guiding analogy lay in crystallography.[94] In 1845 he no longer distinguished between "hemisomorphism" and ordinary mineral isomorphism; rather he proceeded to redefine the latter so as to include the former:

Two substances are regarded as [being] isomorphic when their crystals have virtually the same angles ... *and when they appertain to the same crystalline type*.

I am going to modify this definition and I say that two crystals are isomorphic when their axes are sensibly equal and sensibly inclined by the same amount, *whatever be the type to which each of the crystals belongs*.[95]

Before 1845, Laurent had also taken his comparisons of the chemical and crystallographical natures of his derivatives in another direction—isomerism:

I suppose that a substance C*H ... HH, treated first with chlorine and then bromine, yields the following compound: C*H ... BrCl; what would happen if one treated it first with bromine and then with chlorine? Will the new compound be isomeric or identical with the preceding one? And in the case where it will be isomeric, will it also be isomorphic with it?[96]

For such derivatives, which he claimed to have obtained by treating naphthalene with the halogens as he had outlined, Laurent coined the term, *isomerimorphs*, to denote their combined isomerism and isomorphism.

Finally, by 1845, Laurent had attempted to extend his crystallographical-chemical studies to inorganic mineral substances. "Extend" is perhaps too simple a description for someone like Laurent, whose ideas were always criss-crossing the organic-inorganic interface.

---

[92] A. Laurent, "Nouvelles recherches sur les rapports qui existent entre la constitution des corps et leurs formes cristallines, sur l'isoméromorphisme et sur l'hémimorphisme," *Paris, Comptes rendus* 15 (1842): p. 351.

[93] *Ibid.*, p. 352. *Cf. Revue scientifique*, 14 (1843): p. 81 ff. for more detailed and comprehensive discussion of hemimorphism, extending it to calcite and aragonite *via* a series from calcite to aragonite including barium strontium and lead carbonate. *Ibid.*, p. 83.

[94] A. Laurent, *Paris, Comptes rendus* 19 (1844): p. 1091 ff. N.B. the context for this in plant organography.

[95] A. Laurent, "Sur l'isomorphisme et sur les types cristallins," *Paris, Comptes rendus* 20 (1845): p. 358. Laurent applied this both to mineral series and to the halogenated napthalene compounds with which he had originally taken up the issue. *Ibid.*, pp. 360–366.

N.B. his "planetary" dynamical model for what he called "hemimorphes": substances differing in their interfacial angles by somewhat greater amounts than did isomerimorphs, and whose synoptic formulas were not of the same type, and for substances with different synoptic formulas but isomorphic crystal forms. "Recherches sur le naphthum," *ibid.* 15 (1842): pp. 739–745; *Revue scientifique* 14 (1843): p. 92 (in context here of argument that the fundamental radical unites directly with a halogen rather than its acid); *Méthode de chimie*, pp. 156–164.

[96] A. Laurent, *Paris, Comptes rendus* 15 (1842): p. 350. He said he had raised the issue in a paper submitted to the Académie the year before.

But basically, the development of Laurent's crystallographical-chemical approach can be schematized as having its origin in inorganic crystallography, its major development and application in organic chemistry, and its extension back into inorganic chemistry. His main concern now was to apply the concept of the chemical nucleus or type and its crystallographical concomitants to inorganic as well as organic substances. It appears to have been in connection with this program that Laurent first gave general consideration to the subject of atomic divisibility—which led him to the adaptation of the gas hypothesis. In the *Méthode de chimie*, his posthumously published textbook, Laurent gave an account of how this had come about:

> When I endeavored to arrange the salts of the sesquioxides, in a classification of which I gave the first sketch some fifteen years ago, [?] I met with a difficulty which arrested me for a long time. I sought to include all salts of the same acid in the same formula. The majority of salts yielded pretty well to my notions, with the exception of the salts of sesquioxides and certain double salts....
>
> But seeing the formulae of ferrous sulphate and ferric sulphate thus written—Fe^2SO^4 and $Fe^4S^3O^{12}$—I could not conceive how the two salts could both be sulphates, for containing such different numbers of atoms, they ought not to have the same atomic arrangement. It was then that I asked myself whether the atoms of chemists might not be divisible.[97]

In fact, in his first published consideration of this question in 1843, he was considerably more ambitious than his comments in the *Méthode* might suggest. For he had hopes of solving a host of the most pressing chemical problems of his time through the concept of atomic divisibility. Among these were two especially relevant to his inorganic structuralism: (1) why were there discrepancies between the equivalent weights of the same element in different series of compounds? (e.g., ferrous and ferric compounds); (2) could the concept of atomic divisibility be of use in extending his ideas on isomorphic substitution (as developed for organic chemistry) to mineral substances? He hoped to explain these as well as other questions in terms of the arrangements of sub-atomic particles he called "elements."[98]

His ideas were well illustrated by his grapplings with the first of the above questions, concerning what would subsequently be called polyvalent states, such as the series of compounds formed by manganese or mercury and manifested by multiple equivalent weights of the element. In the case of manganese, for example, Laurent suggested that the manganese "atom" was in fact a complex of twenty-four manganese "elements": sub-groups of four, six, and eight of which formed the different chemical "units" of manganese which participated in chemical reactions. For mercurous and mercuric salts, Laurent's reasoning was even more revealing, for he made explicit analogy with the organic hydrocarbon radicals:

> And since the condensation and the difference in arrangement of the atoms of carbon and hydrogen give different compounds, one ought to conclude that the difference in the condensation and arrangement of the mercurial elements in calomel and corrosive sublimate gives two different metals.[99]

For complex isomorphic minerals in which the isomorphic components could be present in variable proportions, Laurent thought that his ideas on divisible atoms yielded light and simplicity. Rather than giving the usual complex formulas for such minerals, one could devise a *type* formula in which the isomorphic components would be made up of whatever relative proportions of their "elements" the gravimetric data required. Again, Laurent was very much concerned with the measurement of crystal angles.

An example of what he had in mind from his paper of 1843, was his interpretation of the chemical structure of the mineral dolomite, a calcium-magnesium carbonate which crystallized in rhombohedra with interfacial angle of 106°. Calcite (the calcium carbonate) and giobertite (the magnesium carbonate) also crystallized in rhombohedra, with interfacial angles respectively of 105° and 107°. In Laurent's explanation, there was a general isomorphic type for these three minerals which could be characterized by the formula $C^2O^3(c^x m^y)$, where c and m were single "elements" of calcium and magnesium (the combined total of c + m elements always equalling twenty-four). The formula for dolomite would thus be $C^2O^3(c^{12}m^{12})$, expressing its intermediate position between calcite and giobertite.[100]

The notion of atomic divisibility, although elaborated by Laurent first for use in mineral chemistry, played an important role in sensitizing him to the gas hypothesis. Indeed, one of his earliest references to Ampère on atomic divisibility was in connection with divisibility of metallic atoms, in answer to an objection of Baudrimont. In his important paper on nitrogen compounds of 1846, Laurent extended this concept to gaseous molecules, implementing the suggestion of his friend and collaborator, Charles Gerhardt, that atomic weight and

---

[97] Translation is from: *Chemical Method*, trans. W. Odling (London, Printed for Cavendish Society, 1855), pp. 100–101. It has recently been suggested that Berzelius's ideas on allotropy may have influenced Laurent, but there is no evidence.

[98] A. Laurent, *Revue scientifique* 14 (1843): pp. 102–103.

[99] *Ibid.*, p. 101. *A propos* of the unitary theory of the elements, Laurent wrote here: "Mon opinion est moyenne entre cele des chimistes qui admettent une cinquantaine des corps simples formés d'atomes indivisibles, et celle des chimistes qui regardent ces cinquante corps simples comme de l'hydrogène à différents degrés de condensation." *Ibid.*, pp. 101–102. He did not accept the possibility of transmutation of elements but thought that different valence states (as they would later be termed) were due to the divisibility of atoms. *Ibid.*, p. 102.

[100] *Ibid.*, p. 98.

molecular formulas could be clarified by using uniform volumetric measurements.

In the realm of mineral chemistry and crystallography, Laurent continued developing his ideas, especially in the late 1840's. His program, in dealing with silicates, borates, tungstates, etc., was basically the same: to show that all the complex formulas in each of these classes of compounds could be reduced to one or a few basic types, each type being made up of a radical metalloid "acid" (oxide) pictured as first combined with water (as an addition product) and then, with the H^2 of the water replaced by various metallic atoms or groups of "elements."[101]

This attempt to apply the type theory to complex minerals was not nearly as successful or significant for the development of inorganic chemistry as was his speculation in organic chemistry from whence this was derived; it was not until the twentieth century that these mineral substances could be analyzed in the kind of crystallographical-chemical terms Laurent had in mind. But this speculation occurred at a singularly propitious time, when Laurent came into contact with the young chemist, Louis Pasteur.[102]

There will be more to say in our concluding chapter about the relation of Laurent's work to Pasteur's early studies on enantiomorphism, the subject of his spectacular early success. For the present, it is sufficient to point out that Pasteur's first chemical researches, directed in fact, by Laurent, were very much in the vein of Laurent's program. They dealt with an attempt at yet further extension of Laurent's "hemisomorphism" concept to actual dimorphic substances in order to show that they were really almost isomorphic. Laurent, in his turn, used Pasteur's conclusions as support for his most general contention, that all metallic oxides and sulfides were isomorphic.[103]

So far, the discussion of a crystallographical-chemical approach stemming from Ampère has emphasized the chemical side. This was only in reflection of the scientists with whom we have been dealing. While all of them took as paradigmatic the polyhedral molecular unit of Haüy's theory of crystal structure, none of them were particularly concerned with the refinement of the theory *per se*. Rather, they used it and crystallographical data primarily for the elucidation of chemical issues.

The last scientist to be considered as exemplifying this approach was different from the others in that he was a crystallographer by training and profession, and hence, his interest was centered mainly upon the refinement of the molecular theory of crystal structure. This was the crystallographer and mineralogist Gabriel Delafosse, one-time pupil of Haüy and subsequently Pasteur's teacher of crystallography. It was not simply these biographical accidents, but also his particular crystallographical interests in the 1840's, that made Delafosse the most direct link between Haüy's crystallography and Pasteur's work on enantiomorphism. Since Delafosse's work also served as the basis for the geometrical theory of crystal lattice structure worked out contemporaneously with Pasteur's early researches, by Auguste Bravais, he occupies a particularly strategic position in the history of the development of theories of material structure. Moreover, the details of Delafosse's ideas were closely related to those of both Ampère and Laurent.

Delafosse had been closely associated with his teacher Haüy; he was something of the latter's protégé and aide from the time he was graduated from the École Normale in 1816 until Haüy's death. He played an important, not to say even crucial, role in seeing into publication his aged mentor's *Traité de cristallographie* and the second edition of his *Traité de minéralogie*. He was connected with the Muséum d'Histoire Naturelle until 1840; he also taught at the École Normale from 1826 and was named *professeur titulaire* of mineralogy in the Faculté des Sciences and professor at the Sorbonne in 1840.[104]

Beginning in 1840, Delafosse produced a series of memoirs on the refinement of the theory of crystal structure in which he set forth a program of extending Haüy's theory, which he conceived of as a physical rather than formal geometrical one, down to the atomic level, i.e., to the description of atomic arrangement within the crystalline molecule. In this, Delafosse was conscious of the precedent of Ampère. He wrote of what he called the "principle of Ampère . . . which has been up to now and will always be the point of departure of researches dealing with the relations of form and composition. I have always derived this principle from that celebrated physicist."[105] Somewhat later he wrote: "Ampère was the first to attempt to determine the atomic proportions of combinations according to certain polyhedral forms which he regarded as being the representative forms of their molecules."[106] But, unlike

---

[101] A. Laurent, "Sur les silicats," *Paris, Comptes rendus* 23 (1846): pp. 1055–1058; "Sur les borates," *ibid.*, 24 (1847): pp. 94–95; "Recherches sur les tungstates," *ibid.* 25 (1847): pp. 538–543.

[102] The paper on tungstates figured prominently in Pasteur's chemistry thesis. *Cf.* below, chap. 6.

[103] A. Laurent, "Sur l'isomorphormisme [sic] des oxydes RO et R²O³, et sur l'hemimorphisme," *Paris, Comptes rendus* 27 (1848): pp. 134–138.

[104] *Notice sur les travaux scientifiques de M. Delafosse* (Paris, E. Thunot, 1851), pp. 1–2. *Cf.* R. J. Haüy, *Traité de minéralogie* 1 (1822): p. xlvii and *Traité de cristallographie* 1: pp. xl and footnote 1 for his credits to Delafosse.

[105] G. Delafosse, "Réponse à une réclamation de priorité soulevée à l'occasion du mémoire qu'il lu à l'Académie le 17 janvier 1848: par M. Baudrimont," *Paris, Comptes rendus* 24 (1848): p. 336. Baudrimont's reclamation had appeared in *ibid.*, pp. 209–212. Delafosse went on to modify Ampère's ideas for his own purposes.

[106] G. Delafosse, *Paris, Mém. savans étrang.* 13 (1852): p. 546.

Ampère's application of analogies culled from Haüy's theory to the working out of a geometrical chemistry, Delafosse's aim was to elucidate molecular form and atomic arrangement directly from the study of crystal structure.

In 1843 Delafosse proceeded to implement this program.[107] He felt that Haüy had not really succeeded in penetrating to the true molecular level in his theory. In order to refine Haüy's theory and render it more truthful, Delafosse suggested two modifications which were to have far-reaching importance. First, he distinguished between Haüy's *molécule intégrante* and the true physical and chemical molecules which composed crystals. The former were, in Delafosse's refinement, nothing but the reticulate arrangement of the latter in the crystal.[108] In so redefining Haüy's *molécule intégrante*, Delafosse laid the basis for the theory of crystal lattice structure which Bravais was to work out at the end of the decade.

But, though Delafosse was unconcerned with the forms of the real molecules for purposes of envisaging his reticulate crystal structure, he by no means thought of them as being merely abstract points nor did he consider the analysis of crystal structure to be simply an exercise in geometry. Indeed, it was precisely molecular form and structure that Delafosse hoped to get at. In connection with this goal, he made his second modification to Haüy's theory by trying to refine the exact relationship between the formal properties of the macroscopic crystals, and those one inferred for its molecules. It was through his work in this aspect of the theory of crystal structure that Delafosse influenced Pasteur.

Although his first modification had consisted in distinguishing his crystalline molecule from Haüy's *molécules intégrantes*, Delafosse, like Haüy, considered them to be small polyhedra, all orientated in the same direction within the crystal and occupying the nodes of the crystalline network. However, whereas Haüy had essentially read off the form of the *molécule intégrante* from the general symmetry characteristics of the crystal, Delafosse thought that this was not enough for all cases. There were crystals possessing certain special formal and physical properties which seemed to Delafosse to require more attention than Haüy had given to them. For he thought that these properties offered particularly good insight into molecular form and structure.

Delafosse concentrated on a set of properties of this sort with which he felt Haüy had been particularly unsuccessful. These belonged to the so-called hemihedral crystals: crystals with incomplete symmetry, in which Haüy's own principle of symmetry, that any modification on an angle or edge was reproduced on all other symmetrically placed angles and edges, was not fulfilled. Haüy knew about those types of crystals, but he had considered their aberrant forms to be merely superficial and accidental, indicating the play of external forces upon the crystal.[109]

Hemihedry was also associated with certain peculiar physical properties: anomalous surface striation, electrical polarity and optical activity—properties which Delafosse considered as indications that hemihedry was more than superficial. By taking hemihedry and its associated physical properties seriously, Delafosse thought he could advance on to true molecular form and structure.

The case of boracite crystals, to which Delafosse gave the most attention, illustrated his approach. Boracite crystallized in the cubic system and its primitive form seemed to be cubic; Haüy had assigned the cube to its *molécule intégrante*. But the modifications on alternate corner angles of boracite crystals were different: four alternating angles had one kind of modification, the other four, another. Moreover, boracite also exhibited electrical polarity when heated. Finally, boracite crystals (and others like them) exhibited surface striations parallel, not to the edge as was usual, but to the diagonals of the faces.[110]

All of these properties were inexplicable in terms of normal cubic symmetry properties without *ad hoc* qualifications. They were deducible and relatable to each other, Delafosse realized, if one assumed a non-cubic form for the physical crystalline molecules—specifically, the regular tetrahedron. Such tetrahedra, arranged so that their axes were aligned parallel to the axes of electrical polarity (i.e., the diagonal of the crystal) would produce a macroscopic external crystalline form indistinguishable from a cube. But microscopically, the requisite non-symmetry would be established, since each of the axes paralleling the electrical axes would begin in the base of a tetrahedron and end in the *apex* of another (fig. 3). More complex molecular structures, variants on the tetrahedral form, some of which Delafosse suggested, would produce the same macroscopic form.

---

[107] G. Delafosse, "Recherches sur la cristallisation," *Paris, Mém. savans étrang.* 8 (1843): pp. 641–690. This was a reworking of his doctoral thesis presented to the Faculté des Sciences in 1840. A short version of the 1843 paper had been published as: "Recherches relatives à la cristallisation considérée sous les rapports physiques et mathématiques. Ire partie. Sur la structure des cristaux, et sur les phénomènes physiques que en dépendent," *Paris, Comptes rendus* 11 (1840): pp. 394–400.

[108] "... la molécule intégrante d'Haüy n'est que le plus petit des parallelepipedes que forment entre elles les molécules voisines, et dont elles marquent les sommets; ou, s'il on veut, elle n'est que la représentation des petits espaces intermoléculaires ou des mailles du reseau cristallin." *Mém savans étrang.*, p. 650. Delafosse's terms of differentiation were: *molécule intégrante* (or *particule*) vs. *molécule physique*. Ibid., pp. 649, 650.

[109] R. J. Haüy, *Essai d'une théorie . . .*, p. 216, footnote 1. G. Delafosse, *Paris, Mém. savans étrang.* 8 (1843): pp. 655–656.

[110] *Ibid.*, pp. 663–676.

He also took up the molecular forms and structures of such minerals as tourmaline, pyrite, quartz, and beryl, all of which manifested asymmetrical properties and he provided explanations similar to his boracite model.[111] His consideration of quartz was particularly significant, for here Delafosse tried to give an explanation in terms of molecular structure for the property of optical activity. It was this work that led on most directly to Pasteur's.

The 1843 memoir was a milestone in nineteenth-century science, leading on as it did both to Pasteur and Bravais. But it outlined what was only the first stage of Delafosse's more general program. The next stage, moving beyond crystallographical and physical properties to chemical properties, brought Delafosse much nearer to the kind of crystallographical-chemical speculation of Ampère and Laurent. In the 1843 memoir, Delafosse himself stated clearly what he saw as the next stage:

We have seen that by physical and crystallographical considerations alone one can only really determine the genus of the molecular type, that is, a group of simple forms among which this type is very likely comprised. Can one hope to go further, to arrive at the knowledge of the veritable specific type for the molecules of [at least] certain mineral combinations? We believe that someday this will be achieved, but only by combining physical and crystallo-

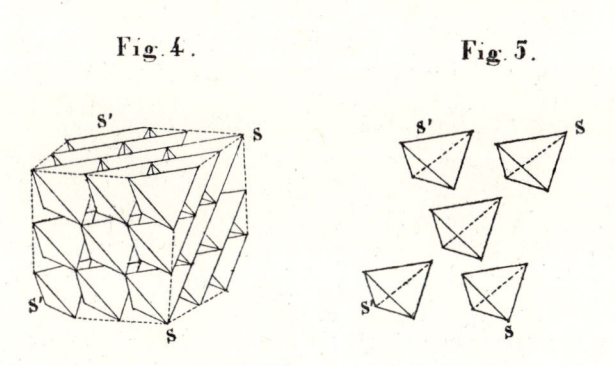

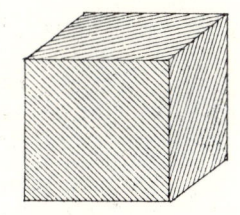

Fig. 3. Delafosse's model of the molecular structure of boracite. "Recherches sur la cristallisation," Paris, Mém. savans étrang.

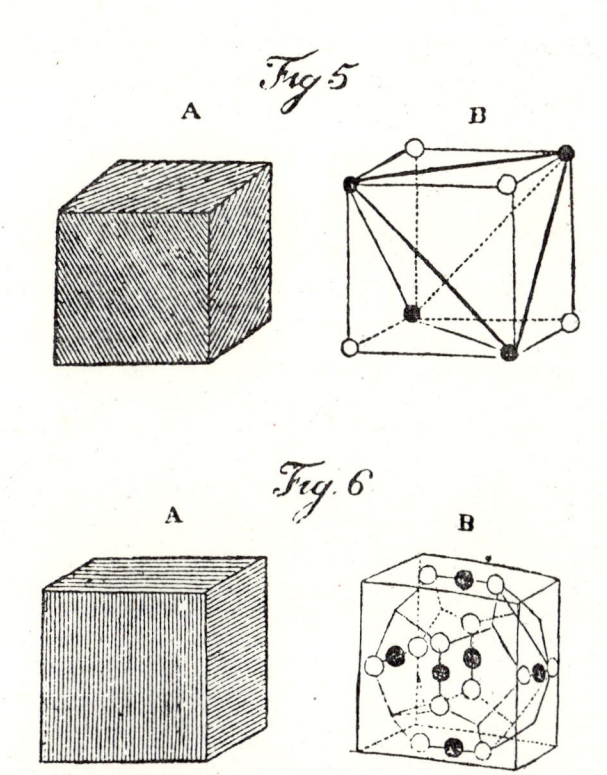

Fig. 4. Delafosse's models of the molecular structure of boracite and iron pyrite. "Crystallographie," Encyclopédie du dix-neuvième siècle.

graphical data with the most unquestionable results of the chemical atomic theory. We have succeeded in constructing geometrically certain atomic formulas, in seeking to reconcile the evidence of crystallography and of chemistry, without doing violence in any way to the accepted ideas of the one or the other science.[112]

One of these tentative models was published by Delafosse in an article on crystallography which he wrote for the Encylopédie du dix-neuvième siècle.[113] It is particularly interesting, aside from its ingenuity, as exemplifying the sort of molecular constructions Delafosse said he had included in his lecture courses, which Pasteur attended. This model was of pyrite crystals, already treated in the 1843 memoir but not in much detail. The primitive form is cubic; the secondary form often a pentagonal dodecahedron. Pyrite had been one of Haüy's most striking triumphs in the derivation of this secondary form using cubic integrant molecules and ingeniously simple laws of decrement.[114]

---

[111] Ibid., pp. 676–690.
[112] Ibid., pp. 675–676.
[113] "Cristallographie," Encyclopédie du dix-neuvième siècle (28 v,. Paris, Bureau de l'Encyclopédie du XIX Siècle, 1842–1859) 9 (1846): pp. 311–324.
[114] Ibid., p. 318. Cf. below, chap. 6, for the appearance of this model (and others) in Pasteur's notebooks. For modern description of pyrite's crystal structure, cf. W. H. and W. L. Bragg (ed.), The Crystalline State 1: W. L. Bragg, A General Survey (London, G. Bell & Sons, 1933), p. 59.

Although pyrite does not show hemihedry in the ordinary formal sense, it is indicated by the curious pattern of striation: on any given face, striations are all parallel to the edges, but striations of two adjacent faces are perpendicular to each other. Taking into account the accepted chemical atomic formula for pyrite, FeS_2, Delafosse devised a model of atomic arrangement to account for all these characteristics: cubic symmetry, dodecahedral secondary form, and the odd striation (fig. 4).

In 1848 (a very significant year for our account, since it saw the publication both of Pasteur's and Bravais's first major studies relating to crystal structure), Delafosse read an ambitious memoir to the Académie des Sciences, in which he sought to implement a portion of the second stage of his program by taking in hand a class of mineral substances, the silicates, which had particularly taxed the ingenuity of chemists and mineralogists by their complexity. In this memoir, Delafosse gave what was in effect something of a synthesis of the whole crystallographical-chemical approach of the first half of the nineteenth century. In its introduction, Delafosse presented a short retrospective survey of the development of this kind of speculation, in which he acknowledged the importance of Ampère's pioneering speculation and reviewed the efforts of Gaudin and Baudrimont, the first very critically, the latter sympathetically.[115]

To deal with the main subject of his memoir, atomic arrangement in the silicates, Delafosse made use of three guiding principles, the first explicity derived from Ampère, the second, from his own modification of Haüy's molecular theory of crystal structure and the third most likely adapted from Laurent. As first principle, Delafosse premised that the atoms which composed molecular polyhedra were "placed such that their centers of gravity always occupy identical apices of the polyhedron which they outline in space."[116] The second had been the guiding principle of his hemihedry paper: "the form of the molecules must always accord with that of the [crystalline] body, consequently, it must be one of the same forms as of its [the crystal's] crystallographical system."[117] Finally, the principle which Delafosse hoped would enable him to penetrate beneath molecular form to true chemical formula and atomic arrangement was like Laurent's fundamental radical model:

In this case, the most important of all for the object which I have in mind in this memoir, the molecule is made up of a central nucleus [noyau] and an exterior envelope, and it is this superficial envelope which determines most immediately the form of the molecular group.[118]

The inorganic paradigm for the employment of this model was hydrated salts such as potassium alum, in which Delafosse conceived the anhydrous salt as the nucleus and the water of hydration as the envelope. Following his third principle, Delafosse tried to relate crystal form to chemical composition through a device he called the "scales of numbers." These expressed the possible numbers of atoms or molecules—molecules of water in the case of the hydrated salt—which could occupy positions surrounding the nucleus compatible with the symmetry of the molecules (as derived from the symmetry of the crystals). Once again, Delafosse hoped to perceive in crystalline symmetry the "scale of numbers" for each alumino-silicate substance.[119]

He thus integrated virtually all the elements of the crystallographical-chemical approach in his speculations on the silicates and, indeed, he passed beyond most of the practitioners of it like Ampère and Laurent by really trying to ascertain atomic arrangement and molecular form, rather than using ideas taken analogically from the theory of crystal structure. But, if this was the climax of a tradition, it was also its denouement. At best, this tradition had been carried on by a small group of rather speculatively inclined scientists, none of whom—even Laurent and Delafosse—positively influenced many of their contemporaries with their specific models and speculations. Nor, perhaps because of the emphasis upon speculation, was there much in the way of development of this approach from one scientist to another. While Laurent was influenced by Baudrimont to turn to a structuralist approach, for example, it would certainly be too much to say that he built directly upon Baudrimont's models of the *Introduction*.

In organic chemistry, the structuralist ideas espoused particularly by Laurent in the 1830's were to be, indirectly, through the complicated formulations and reformulations of the type theory of the 1840's, of great significance in leading to the valence concept and ultimately, structural organic chemistry.[120] In inorganic chemistry the specific applications of the fundamental-radical model was not to lead to anything. Though

---

[115] G. Delafosse, *Paris, Mém. savans étrang.* 13 (1852): pp. 546–549. Two things in particular bothered Delafosse about Gaudin's speculations: Gaudin's overriding concern for rectilinear symmetry caused him to ignore what was to Delafosse the fundamental precept of molecular crystallography—the absolute dependence of crystalline symmetry upon that of its molecules—and Gaudin's thoroughgoing anti-dualism. Delafosse's own molecular structure theory was very similar to Laurent's.

[116] *Ibid.*, p. 550. He went on to say: ". . . et c'est là la seule idée importante que j'emprunte a son système."

[117] *Ibid.*

[118] *Ibid.* Although he made no mention of Laurent here, he did make explicit use of Laurent's widened definition of isomorphism in a paper on plesiomorphism, where minerals crystallized in the same or nearly the same form, although there existed no discernable analogy between their chemical formulas. G. Delafosse, *Mémoire sur la plésiomorphisme des espèces minérales* (Paris, E. Thunot, 1851), especially pp. 31–34.

[119] G. Delafosse, *Paris, Mém. savans étrang.* 13 (1852): p. 550. For silica, he adopted, through an assumption of simplicity, the incorrect formula, SiO.

[120] N. W. Fisher, *op. cit.*, and "Kekulé and Organic Classification," *Ambix* 21 (1974): pp. 29–52.

Laurent and Delafosse could not have known it at the time, far too much was still lacking: valency, atomic size, theories of chemical bonding, and, last but certainly not least, the very geometry of the possible internal atomic arrangement in crystals. Delafosse himself was to write, rather sadly, several years later about the whole Ampère-derived tradition and particularly his own, Laurent's, and Baudrimont's attempts to apply organic-derived models to inorganic chemistry:

> These diverse attempts have been able, in some cases, to furnish an acceptable solution to the problem which it was to resolve; several results thus obtained even presented a fairly high degree of certainty, but they offered no absolute certitude, because one could replace them [the models] with other solutions, of simple theoretical views, with which one can be content until better ones are discovered.[121]

Yet, if the details of these attempts were to prove premature, the general crystallographical-chemical approach, based on the theoretical concept of the polyhedral crystalline molecule, whose formal characteristics were manifested in those of the crystal itself, was to bear positive fruit in the work of Bravais and Pasteur. Bravais, starting in 1848, published a series of epochal papers on the mathematical theory of crystal structure, in which he worked out the fourteen possible space lattice structures.[122] A good deal of the basis for this work was owed to Delafosse's distinction between Haüy's *molécule intégrante* (as representing the reticulate arrangement of the crystalline matter) and the true physical molecule. In his paper on space lattice structure of 1848, Bravais considered molecular structure in a purely abstract manner, as the arrangement of series of points. This abstract, purely mathematical aspect of Bravais's work has been emphasized—I would say overemphasized—in the assessments of his achievement, mostly because subsequent crystallographers in the second half of the ninteenth century who built upon his work considered the theory of crystal structure a branch of mathematics.

But, if Bravais never tried to work out atomic arrangements and molecular forms for specific substances as Delafosse did, he certainly shared with the latter the belief that crystals were made up of molecular polyhedra, placed at the nodes of his space lattices, whose symmetry was intimately linked with that of the macroscopic crystal and whose "preexistent symmetry in the molecular polyhedron is the cause of the symmetry which is established in the crystal, between the positions occupied by the centers of gravity of the molecules."[123] In his *Études cristallographiques* of 1849, Bravais attempted to work out, in a much more general manner than had Delafosse, exactly what were the symmetry relations between these molecular polyhedra and the crystals they formed.[124] Here, he laid down the basis of what eventually was to be the modern determination of atomic and molecular arrangements in crystals.

Bravais was quite conscious of the background of his own studies in the *Études cristallographiques,* in the speculations starting with Ampère and in the work of Delafosse:

> M. Bravais recalled that previously Ampère, in 1814, had arrived, by considerations of a different nature, at the result "that the molecular polyhedron must be formed out of atoms arranged symmetrically around centers of gravity," and that [also] previously, in 1840, M. Delafosse had attributed hemihedry to differences in the structure of the molecule, nevertheless without making known the general rules suited to enable us to pass from the knowledge of the [crystalline] faces suppressed by hemihedry to the determination of the figure of the molecule, or *vice versa*.[125]

An even more direct outcome of the crystallographical-chemical tradition was Pasteur's discovery of the relationship between the crystallographical, chemical, and optical properties of tartrate and paratartrate crystals. We shall turn to these presently, but before the full extent of his achievement can be understood, attention must be given to the developments in another science, physical optics, in the first half of the nineteenth century, developments which also seemed to offer insight into the formal and material nature of crystals.

V. PHYSICAL OPTICS AND THE MOLECULE

The early decades of the nineteenth century, so important for the history of chemistry and crystallography, also witnessed the watershed in the development of physical optics, the "revolution" in theories about the nature of light and luminiferous propagation associated with the struggle between the undulatory theory of Thomas Young and Augustin Fresnel and the hitherto dominant corpuscular theory. One of the foci of optical investigations was the set of phenomena in the crystalline medium associated with double refraction. Double refraction was first noted in 1668 by Erasmus Bartholin in crystals of Iceland spar. In this phenomenon, an incident beam of light undergoes not one but two refractions in traversing a crystal: an ordinary one, following the sine law for the medium, and an "extraordinary" one as well. Thus, an individual image when viewed through the crystal, appears to be doubled. Double refraction attracted the attention both of Huyghens and Newton in the later part of the seventeenth century, be-

---

[121] G. Delafosse, "Rapport sur les progrès de la minéralogie," *Recueil de rapports sur les progrès des lettres et des sciences en France* (Paris, L'Imprimérie Impériale, 1867), p. 68.

[122] Major ones collected as *Études cristallographiques* (Paris, Gauthier-Villars, 1866).

[123] A. Bravais, "Cristallographie," Séance du 19 mai 1849, *Paris, Soc. Philom. proc. verb.* (1849): pp. 51–52.

[124] "Études cristallographiques" in *Études cristallographiques,* pp. 101–276.

[125] A Bravais, *Paris. Soc. Philom. proc. verb.* (1849): p. 52. *Cf.* also *Études cristallographiques,* p. 205.

coming important both to Huyghens's early version of the undulatory theory of light and to Newton's corpuscular theory.[1]

Through the work first of Etienne Malus and subsequently of François Arago, Jean Baptiste Biot, David Brewster, and others, the phenomenon of double refraction was widened out in the early nineteenth century to include a host of related phenomena not only in crystals —all termed "polarization" by Malus. These seemed capable of offering insight, perhaps the most penetrating insight of all, into crystalline and molecular structure and composition. In this context, physical optics came into interaction with crystallography and chemistry.

Aside from its peculiarity and relative rarity, double refraction was interesting because the observed effects were related to the orientation of the incident ray and the observer to the crystal. For example, Huyghens noted that the incident and doubly refracted rays were all in the same plane only when the incident ray, and the normal to the crystal face at the point of entry of this ray, were themselves in a plane which Huyghens called the "principal section" or in planes parallel to it. In the case of the Iceland spar rhombohedron, the principal section was the plane normal to a face and passing through the major axis of the crystal.

On the basis of the geometry of double refraction along the principal sections of all the faces of doubly refracting crystals, Huyghens was able to construct a model for double refraction in terms of his undulatory theory of light propagation which also highlighted the relationship of the phenomena to crystal orientation. In it, the ordinary refraction was propagated in the crystal by spherical waves; the extraordinary refraction, however, through spheroidal waves whose axis was parallel to the axis of the crystal in the case of a rhombohedron. Huyghens also demonstrated geometrically that the double refraction would reduce to single, ordinary refraction along the crystal axis.[2]

Even more striking was the relationship of crystal orientation which Huyghens noted when a light ray was passed through two double refracting crystals in succession. When the two crystals were aligned with their principal sections parallel, the ordinary and extraordinary refractions which resulted from a ray of light traversing the first crystal suffered no new double refraction when they passed through the second. Rather, each ray was again refracted in the manner it had been initially, the ordinary ray being again refracted according to the ordinary law of refraction and the extraordinary ray following its special refraction. However, if the second crystal were then rotated with respect to the first until their principal sections were perpendicular to each other, the following was observed: as the crystals changed their relative positions from parallel to oblique, the double refractions of the first crystal began to produce new double refractions in the second, so that four rays in all appeared after the traversal of both crystals. The two new rays, faint at first, gradually brightened as the rotation continued, while the initial two rays dimmed. When the crystals became perpendicular to each other, the initial continuations of the ordinary and extraordinary rays had disappeared and the new rays remained instead, so that at this position the ordinary refraction of the first crystal gave rise to one extraordinary ray in the second and the initial extraordinary refraction produced an ordinary one upon its successive passage. If the rotation were then continued past perpendicular until the crystals were again parallel, the reverse sequence occurred: new double refractions at the oblique positions, gradually brightening with the dimming and final extinction of the rays observed at perpendicular so that at the new parallel positions, the original refractions were observed again. Huyghens, though describing this phenomenon in detail, was unable to provide an explanation for it in terms of his undulatory theory,[3] and Newton was to use it as his principal argument against Huyghens's theory.[4]

It was this second phenomenon which was extended in its range and interest by a discovery made by Malus in 1808. He found that light which had been reflected by surfaces at certain angles and then passed through a doubly refracting crystal behaved as if it had *already* been doubly refracted once and were now being passed through a second doubly refracting crystal. Malus's discovery of "polarization," as he called this, was hardly an isolated happening. On the contrary, it was but the most spectacular result of the increasing amount of attention paid to optical phenomena generally and double refraction in particular, in the preceding two decades.

One of the first scientists to resume the study of double refraction in a systematic fashion in this period had been Haüy. Given the obvious relation of this phenomenon to crystal structure, it is hardly suprising that Haüy should have taken up its study. By his own admission, it had been his mineralogical investigations which had directed his attention to double refraction in the first place.[5] His discovery that there was no general agreement among the previous investigators concerning the mathematical function governing the extraordinary refraction led Haüy to make a general reexamination of the measurements and theories of his predecessors,

---

[1] A. I. Sabra, *Theories of Light from Descartes to Newton* (London, Oldbourne, 1967), pp. 221–229.

[2] C. Huyghens, *Traité de la lumière* [Leiden, 1690], *Œuvres complètes de Christiaan Huygens* (22 v., La Haye, Martinus Nijhoff, 1888–1950), 19: pp. 494–521.

[3] *Ibid.*, pp. 517–518.

[4] I. Newton, Queries 25, 26, 28, *Opticks* (2nd ed., English, London, W. and J. Innys, 1718), pp. 328–336, 338–339. *Cf.* A. I. Sabra, *op. cit.*, pp. 226–229.

[5] R. J. Haüy, "Mémoire sur la double refraction," *Paris, Mém. Acad. Sci.* 1788 (1791): p. 35.

particularly Huyghens and Newton, and this was the subject of his first publication on double refraction. Haüy found that his own measurements tended to confirm Huyghens's construction of the paths of the ordinary and extraordinary rays. However, Haüy, like most of his contemporaries, adhered to Newtonian corpuscularism and, therefore, he attempted to derive an alternative construction to that of Huyghens which would not require the undulatory spheroids.[6]

But perhaps even more central to Haüy's interests was the use of double refraction to confirm his own theory of crystal structure. In a paper he published in 1793, the intent of which was to confirm both the forms and the orientations within the crystals of the primitive forms he had worked out for a number of different minerals by testing for the orientations of the principal sections of these primitive forms, Haüy stressed the potential value of double refraction for analysis of the crystal structure:

... if the relationship indicated by the substances already submitted to test are sustained in other minerals, it would offer at once a further guarantee in favor of the theory concerning the structure of crystals, and a guide for penetrating further into the [theory] of [this] phenomenon, in generalizing the cause on which it depends.[7]

In his great textbooks of the early years of the new century, the *Traité de minéralogie* and the *Traité élémentaire de physique*, Haüy devoted considerable attention to double refraction and again stressed its usefulness in confirming his theory of crystal structure. In addition to general surveys of its investigation, Haüy offered some new observations of his own, the most important of which were that crystals possessing "regular" primitive forms—cubes, regular octahedra, and rhombic dodecahedra—apparently exhibited no double refraction, and that light passed down the axis of a rhombohedral Iceland spar of primitive form (incident upon and exiting from the faces produced by truncating the opposite obtuse solid angles) was not doubly refracted.[8]

Thus in his articles and textbooks Haüy had summarized and synthesized the accumulated knowledge about double refraction. He had afforded support to Huyghens's geometrical construction for the phenomena, if not to his undulatory explanation of it. He had extended the range of its study to a fairly large number of mineral substances and, finally, he had made a number of correlations between crystal structure and double refraction which were to prove significant.

Others in the years just after the turn of the century also took up the investigation—or reinvestigation—of double refraction. In 1802 Wollaston published a study confirming Huyghens's construction and supporting Huyghens's interpretation much more positively than had Haüy. Indeed, Wollaston had taken up this investigation at the behest of Thomas Young, the proponent of the wave theory of light.[9] In 1807 Haüy's own observations on double refraction were taken up as a point of departure for a new approach to the relating of this optical phenomenon to crystal structure by the German botanist and naturalist, J. J. Bernhardi.

Like many of his contemporary *Naturphilosophen*, Bernhardi was an opponent of atomic theories of material structure of which Haüy's was, at this time, the most noteworthy example. Bernhardi attempted to devise a simpler alternative to Haüy's primitive forms and decrements and he criticized Haüy's own relating of double refraction to his primitive forms as missing what Bernhardi considered to be the point of this phenomenon: its relation to a particular axis in the doubly refracting crystal which he termed the "light-axis," which connected two opposing and oppositely situated "light-poles." This consideration of material structure in polar terms was common to early nineteenth-century *Naturphilosophie*.

Starting with the usual rhombohedron where the light poles were coincident with the apices of the two obtuse solid angles, and the light-axis with the axis between them, Bernhardi generalized that such poles and axes existed for all nonregular crystal forms. Double refraction was caused by attractive forces emanating from the light pole in whose direction the refractions were inclined and thus, he deduced that there could be no double refraction in those directions of incidence which precluded such inclinations, that is, in planes parallel or perpendicular to the light axis.[10]

In France by 1807 interest in double refraction was sufficiently great for the First Class of the Institut to set the derivation of a mathematical theory for it, with experimental verification, as the topic of an essay to be awarded the mathematical prize in January, 1810. While studying double refraction with this prize in view, Malus made his discovery of polarization in the autumn of 1808.[11] Earlier in 1808, the physicist Laplace had taken up the mathematical investigation of double refraction partly, it seems, through interest generated by some of Malus's pre-polarization work. By this time, enough evidence had accumulated to demonstrate the correctness of Huyghens's construction; the task of the

---

[6] *Ibid.*, pp. 44 ff.

[7] R. J. Haüy, "Sur la double refraction de plusieurs substances minérales," *Annal. de chimie* 17 (1793): p. 156 (whole article pp. 140–156).

[8] R. J. Haüy, *Traité de minéralogie* (1801) 1: pp. 229–235; 2: pp. 38–51, 196–229; *Traité élémentaire de physique* (2nd ed., 2 v., Paris, Courcier, 1806) 2: pp. 334–355 (p. 352 for comment on usefulness for confirming crystal structure theory).

[9] W. H. Wollaston, *Phil. Trans.*, 1802: pp. 381–386.

[10] J. J. Bernhardi, "Beobachtungen über die doppelte Strahlenbrechung einiger Körper," *Gehlen, Jour.* 4 (1807): pp. 252 ff. (pp. 230–257 for whole article).

[11] M. P. Crosland, *The Society of Arcueil* (London, Heinemann, 1967), pp. 342–346. Malus's own previous work had probably contributed to the establishment of the prize.

Newtonian-oriented physicists was to show how this same construction could be derived from Newtonian assumptions of light corpuscles being refracted because of the attraction of material molecules. Laplace succeeded brilliantly here; employing the principle of least action and the assumption that the velocity of the extraordinary ray was augmented by a function of the angle between the incident ray and the shorter axis of rhombohedral crystals, Laplace was able to derive Huyghens's construction in Newtonian terms.[12]

Laplace's dynamical derivation for double refraction, like Bernhardi's, emphasized the importance of the crystalline axis in this phenomenon, but unlike Bernhardi, Laplace was and always had been an adherent of a molecular concept of material structure and of Haüy's theory of crystal structure. To him, then, double refraction, as earlier chemical affinity, seemed capable of offering clues about the nature of the component molecules of crystals. Since these molecules were the source of the forces exerted on the incident light rays, therefore each molecule ought to possess the same optical properties as did the whole crystal.[13]

Malus, following his discovery of polarization-by-reflection, proceeded to investigate comprehensively all phenomena associated with polarization and double refraction. These investigations, for which he received the Institut's prize,[14] were to be of fundamental importance to the controversy of the next decade between the supporters of the wave theory versus the corpuscular theory of light and they were to lead to an outburst of investigative activity into the optical properties of crystals.

But his study also had implications for Haüy's theory of crystal structure and some of the issues which had arisen between crystallography and chemistry. Like Laplace and most other French contemporaries, Malus accepted the idea that crystals were composed of small polyhedral molecules. And like Laplace, Malus felt that the geometry of the extraordinary ray refraction, related as it was to the orientation of the incident ray to the crystalline axis, offered potential insight into molecular form. For the extraordinary refraction seemed to be governed in part by the shape and the orientation of the integrant molecules *vis-à-vis* the incident ray.[15] The relationship between double refraction and molecular crystal structure seemed strengthened in the case of Iceland spar by the fact that the axis of double refraction, where no polarization would be observed, was coincident with the major axis of the rhombohedral primitive form of Iceland spar, though Malus himself was unwilling to speculate on the position of the axes of double refraction in other crystalline substances.[16]

Arguing from such premises about the relationship between double refraction and crystal structure, Malus offered an opinion on the controversy over the nature of calcspar and aragonite. He supported Haüy's contention, against chemists like Berthollet, that they must be different substances because the difference in their double refraction meant that their integrant molecules could hardly be the same.[17] At the same time Malus took issue with Haüy over the structure the latter had posited for aragonite since, unlike calcspar, the axis of double refraction did not coincide with the axis of Haüy's primitive form.[18] Moreover, incidentally, Malus had determined a value for the interfacial angle of calcspar which was over half a degree different from Haüy's value, a difference, though apparently small, nevertheless was serious in its implications against the simplicity of Haüy's decrement laws for this substance.[19]

Malus died tragically young in 1812 but the investigation of double refraction in crystals was continued by his colleagues in the years immediately following his work on polarization, and a number of important discoveries were made. In 1811 François Arago discovered what was later to be known as rotatory polarization in quartz crystals. The following year Jean Baptiste Biot made two further discoveries. First, some pairs of crystals, beryl and quartz for instance, had the ability to neutralize the polarization of one another's light when they were aligned together with their principal sections parallel. For beryl, moreover, the extraordinary ray was deflected more than the ordinary in double refraction, whereas the opposite was true for quartz. On the basis of his dynamical assumptions about the cause of double refraction, Biot named the quartz type "attractive" and the beryl type "repulsive."[20] Second, he discovered what was to be known as biaxiality in crystals. Some crystals gave evidence of two axes of double refraction, rather than none or only one. This characteristic was independently discovered by the Scottish physicist David Brewster and first clearly explained by him.[21]

---

[12] P. S. de Laplace, "Sur le mouvement de la lumière dans les milieux diaphanes," *Mémoires de physique et de chimie de la Société d'Arcueil* (3 v., N. Y., Johnson Reprint Corp., 1967) 2: pp. 111–142. Cf. 113–114 for reference to Malus's role in establishing proof of Huyghens's law. (In future references: *Mém. Soc. Arcueil*.)

[13] *Ibid.*, p. 114.

[14] E. Malus, "Théorie de la double refraction," *Paris, Mém. savans étrang.* 2 (1811): pp. 303–508.

[15] *Ibid.*, pp. 466–478, 491–492.

[16] *Ibid.*, pp. 385, 477–478. Malus also knew and appreciated Bernhardi's work. *Ibid.*, pp. 502–503.

[17] *Ibid.*, pp. 456–464.

[18] *Ibid.*, pp. 455–456.

[19] *Ibid.*, p. 333. Cf. Haüy's reposte in the *Traité de cristallographie* 2: pp. 386–395.

[20] J. B. Biot, "Sur la découverte d'une propriété nouvelle dont jouissent les forces polarisantes de certains cristaux," *Paris, Mém. de l'Inst.* part II (1812) (read 25 Apr., 1814): pp. 19–26; "Addition au mémoire sur les deux genres de polarisation exercés par les cristaux doués de la double refraction," *ibid.* (read 15 May, 1814).

[21] J. B. Biot, "Mémoire sur un nouveau genre d'oscillation que

The discovery of all these optical phenomena in crystals associated with double refraction and polarization played an enormous role in the revolution which was taking place in the theory of the propagation of light. But coming as it did at the same time when important developments had just taken place in crystallography and chemistry, it struck a number of the investigators that the newly discovered optical properties of crystalline matter might provide additional and even more useful methods for unraveling the secrets of the material molecules, than had even the recent advances made by chemists and crystallographers.

This was particularly true of Biot. Throughout his extremely long and active scientific career in which he devoted himself to a wide range of scientific (and extra-scientific) interests, one of his most enduring interests was in testing chemical and crystallographical data by optical methods. The results of his labors here were to be an essential component of Pasteur's synthesis of chemical, crystallographical and optical data, and one can well appreciate how moved Biot was at Pasteur's discovery in 1848, representing, as it did, a real climax to his own more than forty years of work in this subject.

One of Biot's first scientific publications in 1806 had dealt with the correlation of chemical and optical properties done jointly with his younger colleague, François Arago. They had tested the refractive indices of a large number of gaseous substances in the hope that this optical property might be useful in chemical analysis.[22] Their reasoning was orthodox Newtonian: refraction from a rarer to a denser medium was caused by the attraction of light particles in the former by the material corpuscles in the latter medium. Assuming that this luminiferous attractive power was of the same nature as chemical affinity,[23] then each diaphanous elementary substance ought to be characterized by its own refractive index, expressing its particular power of attraction for light.[24]

Biot and Arago suggested that the refractive indices of elementary substances might survive in chemical combination even more than did their chemical affinities. Therefore, the refractive index of a compound ought roughly to be the mean of the refractive indices of each constituent multiplied by its weight proportion in the compound.[25]

Biot was a member of the semi-formal Arcueil group, centering around Laplace and Berthollet and devoted to discussions and experimentation concerning topical issues which ranged through all the physical sciences. In 1809, in the *Mémoires* of this group, Biot published an account of another joint project, this one with the chemist Louis Thenard, on the prominent issue focused in crystallography, the calcspar-aragonite controversy.[26] By this time, the puzzle of these substances had become quite deep; several chemical analyses had been unable to uncover any differences, whereas their crystal forms were clearly incompatible. Biot and Thenard attempted, once again, to introduce the refractive index test as a possibly more sensitive means of establishing chemical differences between these substances which their own chemical analyses had been unable to detect. The test of the refractive indices of calcspar and aragonite was inconclusive, although it pointed to their identity.[27]

But if the refractive indices suggested chemical identity, there was still an additional problem regarding their optical properties, which drew Biot's and Thenard's attention. This was the fact that calcspar was doubly refractive whereas aragonite apparently was not. In order to reconcile this apparent contradiction between chemical identity and differences in the optical properties of the crystals, the two scientists postulated that there must be differences either in the molecular form or the mode of molecular aggregation in the crystal, giving one of the first general definitions of what was later to be known as polymorphism:

From this, it is easy to conceive that [the extraordinary refraction] can be very different for molecules composed chemically in the same way, if the elements which compose these particles are united in a different way, or if the molecules are aggregated and crystallized diversely. This is the case for aragonite and calcspar, and it ought not to be surprising that although formed of the same elements, these two substances act differently on light in double refraction, while their action remains the same in ordinary refraction or when their elements are disunited.[28]

---

les molécules de la lumière éprouvent en traversant certains cristaux," *Paris, Mém. de l'Inst.*, part I (1812): pp. 346 ff.; D. Brewster, "On the Affections of Light Transmitted through Crystallized Bodies," *Phil. Trans.* 1814: pp. 202–218.

[22] J. B. Biot and D. F. Arago, "Mémoire sur les affinités des corps pour la lumière, et particulièrement sur les forces réfringentes des différens gaz," *Paris, Mém. de l'Inst.*, 1806: pp. 301–387.

[23] *Ibid.*, p. 302. In a later paper, Biot wrote: ". . . et qu'y a-t-il de plus analogue, je dirais presque de plus identique, que l'affinité chimique est l'action des corps sur la lumière." "Mémoire sur de nouveaux rapports qui existent entre la reflexion et la polarisation de la lumière par les corps cristallisés," *Paris, Mém. de l'Inst.* part. I (1811) (read 1 June, 1812): p. 211.

[24] J. B. Biot and D. F. Arago, *Paris, Mém. de l'Inst.*, 1806: pp. 326–328.

[25] *Ibid.*, pp. 329–330. They found that the "condensation" of the constituents into the products seemed to make a slight, though not serious, difference and there was the more serious anomaly of diamond, whose refractive index was much too high for pure carbon, as deduced from that of other carbon compounds, *ibid.*, pp. 332 ff. *Cf.* also p. 344 for their argument that the fairly close experimental approximation of their principle favored the emission over the undulatory theory of light.

[26] L. Thenard and J. B. Biot, "Mémoire sur l'analyse comparée de l'arragonite et du carbonate de chaux rhomboidal, avec des expériences sur l'action qui ces substances exercent sur la lumière" (read at the Institute, 14 Sept., 1807) *Mém. Soc. Arcueil* 2: pp. 176–206.

[27] *Ibid.*, p. 202. For their conclusion on the chemical identity of these substances from their own analyses, *cf.* pp. 195–196.

[28] *Ibid.*, p. 205.

The subsequent discoveries of other polarization and double refraction phenomena reinforced Biot's belief that optical properties, particularly double refraction, offered clues about the intimate molecular constitution of matter. His work and views on this subject were synthesized in a paper read and published in 1818.[29] The most interesting feature of this paper was Biot's chemical-mineralogical interest and orientation. Through the use of optical crystalline structure—the number of double refraction axes—Biot hoped to provide a means of checking on chemical composition. In particular, he wanted to be able to determine whether or not all the products revealed by chemical analysis were really in combination in the mineral or were mechanically intermixed.

His chemical orientation, which was to remain with him, was Bertholletian: he supported Berthollet in his opposition to the doctrine of definite proportions. By the time of the appearance of this paper, Daltonian stoichiometry was achieving ascendancy on the Continent. In mineralogy the fixed species doctrine of Haüy was also dominant, in France at least, although the controversy over aragonite-calcspar still continued and at the time Biot's paper appeared Beudant had just published his own studies on what seemed to be anomalous crystallizations in variable proportions which were shortly to be subsumed under the rubric of isomorphism.[30]

In both Daltonian chemistry and Haüy's mineralogy, the definition of true chemical combination was based on definite proportions. Biot took issue with this doctrine, at least for natural products. In echo of Berthollet, he suggested that alloys, solutions, earthy minerals, etc., might well be chemical compounds despite their variable proportions, and he appealed to the criteria of chemical combination which Berthollet had invoked earlier in his controversy with the same two supporters of definite proportions. Alloys manifested different properties from their constituent metals; solutions clarified and became transparent, meanwhile taking up or giving off sensible amounts of heat; glasses were transparent and refracted light with perfect regularity.[31]

Moreover, with regard to Beudant's work, neither Dalton's or Haüy's criteria for chemical combination seemed to give much help in establishing whether the various metallic sulfates in Beudant's crystals were truly united or merely intermixed. But Biot believed that optical double refraction structure gave such a criterion.[32] The fact that double refraction was related to clearly oriented axes at every point in the crystal[33] and that it disappeared with the destruction of crystalline structure, convinced Biot that it indicated a regular arrangement of molecules in the crystal, an indication especially useful where crystal form was difficult to observe as in the case of mica.

Despite his differences with Haüy's stoichiometric views, the principle which Biot assumed in his use of double refraction structure as a guide to chemical composition was parallel to that which Haüy had employed in his use of crystal structure as a criterion for resolving this chemical issue. Distinct double refraction structure (regular axes of double refraction) indicated the presence of the true crystalline state, and the particular quality of the double refraction was somehow a function of the true chemical components of the crystal, just as the primitive form had been for Haüy.

On this basis, Biot suggested two "propositions" as general tests of chemical union in crystals: (1) If the presence (and absence) of a component made no difference to the nature of the double refraction structure of mineral samples, that component was probably not chemically united in the mineral. On the other hand, if the addition of a new substance did cause a change in the double refraction structure, it meant that this substance had entered into intimate union with the original constituents. (2) Two substance which exhibited distinct but different optical structures "differ either in their chemical composition, their mode of aggregation, or in both these qualities together."[34]

Biot had hoped to apply these criteria to Beudant's sulfate crystals, but in the absence of good crystal specimens, he turned instead to the mica family,[35] where he classified these substances into four distinct classes according to whether they were uniaxial or biaxial. Biot also saw in the micas a strong counter-example to the law of definite proportions, at least as it pertained to natural mineral products. Here were minerals having the attributes of true chemical homogeneity—transparency, crystal form, and double refraction structure.

---

[29] J. B. Biot, "Mémoire sur l'utilité des lois de la polarisation de la lumière, pour reconnaitre l'état de cristallisation, et de combinaison dans un grand nombre de cas où le système cristallin n'est pas immédiatement observable," *Paris, Mém. Acad. Sci.* **1** (1816): pp. 275–346 (Read 22 June, 1818).

[30] *Ibid.*, pp. 275–280 for his discussion of the calcspar-aragonite controversy, and pp. 280–282 for Beudant's work.

[31] *Ibid.*, pp. 282–285. Biot also cited Gay-Lussac's redaction of the memoir of Bucholz and Meusner, "Expériences pour déterminer la quantité de strontiane dans plusieurs espèces d'arragonite," *Annal. de chimie* **2** (1816): pp. 176–182. Here, Gay-Lussac argued that the variable proportions of strontium in aragonite, as well as of carbon in iron, did *not* mean that they were mere accidental intermixtures. *Ibid.*, p. 177, footnote 1. He also made allusion here, to isomorphic crystal growth of mixtures of potassium and ammonium alums. "C'est là une nouvelle cause à ajouter à celles déjà connues qui peuvent troubler les effects de la loi générale des proportions définites." *Ibid.*, p. 178.

[32] J. B. Biot, *Paris, Mém. Acad. Sci.* **1** (1816): pp. 286–287.

[33] As opposed to glass plates, in which David Brewster had succeeded in developing optical axes. *Ibid.*, pp. 292–293.

[34] *Ibid.*, pp. 295–296.

[35] Because of their lamellar structure, micas were difficult to study crystallographically, although easy optically. In addition, chemical analysis yielded variation in the proportions of the constituents—and even diversity. *Ibid.*, pp. 296–300.

Yet all were composed of widely varying proportions of silica, alumina, magnesia, etc. Therefore, Biot concluded, neither chemistry nor crystallographical mineralogy, with their basis in definite proportions, had general application in Nature.[36] In order to obtain an accurate knowledge of molecular composition, above and beyond the raw results of chemical analysis, it was necessary to turn to optics.

By 1818, Biot had also investigated rotatory polarization,[37] another phenomenon which seemed to be even more promising in affording insights into the nature of individual molecules than did double refraction. It was on this subject that Biot was to concentrate his optical studies thenceforth and his investigations of this phenomenon were to be of decisive influence on the young Pasteur. The research on rotatory polarization up to Pasteur progressed in a kind of dialectical fashion, from the study of crystals, to other material states and then back to crystals. Discovered first in quartz crystals by Arago, it was later found by Biot to be manifested also in organic liquids, solutions, and even gases. The appearance of rotatory polarization in these non-crystalline states particularly excited Biot, since it seemed to him that here rotatory polarization was, of necessity, a property of individual molecules. With Pasteur, the focus of research returned to crystals, but now to the crystals of some of the organic substances whose rotatory polarization Biot had observed in their solutions. The key to Pasteur's achievement lay in correlating this manifestation of rotatory polarization with peculiarities in crystal form.

In 1811, while investigating the effects of certain kinds of mica and of calcium sulfate on polarized light, Arago had noticed an odd phenomenon when polarized light was passed through slices of quartz and then analyzed through a calcspar crystal: if the calcspar crystal was initially aligned with respect to the polarized light ray so that its axis of double refraction was parallel to the plane of polarized light (this could be determined by noticing the orientation at which only the ordinary ray of refraction was visible) and then a slice of quartz was interposed between the calcspar and the source of polarized light with the quartz's optical axis also aligned parallel to the plane of polarization, then no longer would only the ordinary ray of refraction be produced in the calcspar, but rather two images of complementary colors. If the calcspar crystal were rotated, these two images of double refraction went through the entire spectrum of colors in a demi-rotation.[38] Although Arago did not clearly distinguish this phenomenon from the somewhat similar depolarization effect of mica,[39] he did suggest that somehow "the molecules of different colors which make up the two white beams (emerging from the quartz crystal) have their poles directed towards different points in space."[40]

Biot, who was closely following Arago's research (and competing with him) took up the investigation of the effect of quartz on polarized light and quickly recognized its uniqueness and interest. Using thinner sections of quartz than Arago had employed, he was able to modify and extend the latter's observations. For example, he noted that when a quartz slice of only 0.4 millimeter thickness was used and the calcspar crystal was aligned with its optical axis parallel to the plane of the incident polarized light, double refraction would take place, the color of the ordinary ray being white and that of the extraordinary ray a "somber blue." When the calcspar rhombohedron was rotated to his left, the blue extraordinary ray became tinged with violet and, at an angle of 9°45′ with its original position, this ray became invisible. If the rotation was continued, the extraordinary ray reappeared, brightened and passed from red through orange and yellow until it was finally white.

Meanwhile the ordinary ray—initially white—remained white for some 40° of rotation further to the left of the 9°45′ where the extraordinary ray had become nearly invisible, and then gradually took on bluish tones, which deepened as its intensity lessened. At 9°45′ past a quarter revolution leftwards, the ordinary ray became an almost invisible bluish-violet, while the extraordinary ray was now white and bright. As the calcspar rhombohedron was rotated through the entire 360°, this alteration took place three more times, at 90° intervals.

When quartz slices of greater thickness were used, the intial colors of the ordinary and extraordinary rays were different from those of the 0.4 millimeter case. But as the calcspar crystal was rotated, at certain angles the extraordinary ray almost disappeared, assuming the characteristic dull blue-violet color. These angles were greater than the 9°45′ of the first case; they seemed proportional to the thickness of the quartz slice.[41]

---

[36] *Ibid.*, pp. 339–342.

[37] He began another synthetical paper of 1818—on rotary polarization—with this theme: "Dans l'état actuel de la chimie et de la physique, aucunes recherches ne semblent devoir être plus utile et plus fécondes que celles qui concernent les propriétés individuelles des molécules dont les corps sont composés. *Paris, Mém. Acad. Sci.* 2 (1817): p. 41.

[38] D. F. Arago, "Mémoire sur une modification remarquable qu'éprouvent les rayons lumineux dans leur passage à travers certains corps diaphanes," *Paris, Mém. de l'Inst.* part I (1811): pp. 115 ff.

[39] The depolarizing effect of mica (or calcium sulfate): light which had been polarized by reflection and whose plane of polarization was parallel or perpendicular to the principal section of a calcspar analyser—when *first* passed through a mica sheet and then through the calcspar—gave rise to two images (rather than one, as would be expected), but they were of complementary colors. At quarter turns of the mica (or of the calcspar), they coalesced into one white image. *Ibid.*, pp. 6–16 (the numbering is askew here; this corresponds to 98–108 in preceding pagination).

[40] *Ibid.*, p. 122.

[41] J. B. Biot, *Paris, Mém. de l'Inst.* part I (1812): pp. 218 ff.

Biot realized that this phenomenon could be explained if one assumed a rotation of the original plane of the polarized light as it passed through the quartz. In his Newtonian terms, the light molecules acquired a "movement of rotation about their center of gravity" in traversing the quartz crystal slice. Two factors seemed to govern the extent of rotation: the thickness of the quartz section and the part of the spectrum to which the light belonged, with rotation greatest for the violet end and least for the red.[42]

Biot also noticed another striking phenomenon. Some quartz crystals rotated light molecules to the right, that is, the calcspar rhombohedral analyzer had to be rotated slightly to the right rather than the left in order to reduce the extraordinary ray to minimum intensity. The rotation was, moreover, the same as was necessary to extinguish the extraordinary ray for the same thickness of quartz which deviated the plane of polarization to the left; and the superposition of a left-deviating and a right-deviating quartz crystal resulted in an effect on the plane of polarization of the light passed through them which was equal to the vectorial sum of their rotations.[43]

In the next few years, while working on double refraction, Biot continued to experiment on this rotatory polarization, summarizing his results in a paper which he submitted to the Académie des Sciences in September, 1818. In the course of his experiments, he had succeeded in making the important generalization that for a given thickness of quartz, the amount of rotation of the polarization plane for different colors was inversely proportional to what was later called their wave lengths. From this, he was able to show that the division of polarized white light into the differently colored ordinary and extraordinary rays which appeared after the light had passed through the calcspar analyzer, could be accounted for by applying the general law of double refraction to each color component separately.[44]

Beyond these extensions and refinements to the study of rotatory polarization in quartz crystals, Biot had discovered the same power in non-crystalline media: in certain organic liquids, vapors, and solutions, including oil of turpentine, cane sugar solution (which had a left-ward deviation), as well as beet sugar and natural camphor in alcohol, which rotated the plane to the right. He was able to show that the same law of rotation held for these substances which held for quartz, and that the amount of rotation in the liquids was proportional "to the number of particles which the essence of turpentine contained." He demonstrated this by means of an experiment in which tubes of different lengths but the same thickness were filled with the same weight of oil of turpentine, the oil of turpentine in the longer tube being diluted with optically neutral sulfuric ether. When polarized light was passed through each tube, the rotatory effect on the light was the same. And it appeared to be carried over with the turpentine molecules when they entered into combination with other substances, so long as the molecules remained "intact" in their new combinations. Biot also found that vaporized oil of turpentine had optical rotatory power.[45]

This extension of rotatory polarization to liquids and vapors had a crucial bearing on Biot's theorizing about the phenomenon. It convinced him that the power of these substances, at least, appertained to the individual molecules, for there could be no question of an aggregative order among liquid or gaseous molecules which might give rise to it. As he put it, if vapors rotated the plane of polarized light, this would show ". . . that this property belonged to its [the vapor's] particles and, moreover, that these particles do not change their form at all in vaporizing." [46] The positive result of the test of this led to the conclusion:

This last and important experiment . . . succeeded in demonstrating that the singular power which certain substances possess for turning the axes of polarization of light rays, is an individual faculty of their particles, a faculty which they only lose when they cease to be themselves by decomposition.[47]

From 1818 until 1832 Biot produced no new studies on rotatory polarization, having abandoned his optical studies in the 1820's for the excitement of electromagnetism. In this interval, in 1820, an important connection was made between the optical rotatory power of quartz and its crystal form, by the young physicist and astronomer J. W. F. Herschel.[48] The case of quartz, the first substance in which rotatory polarization had been discovered and the only inorganic substance found to possess this power, was different from the organic liquids and vapors Biot had gone on to investigate. For

---

[42] *Ibid.*, p. 256. As a dynamical model, Biot suggested an infinite number of radial axes in a quartz slice, perpendicular to the optical axis, and whose force rotated the light molecules. *Ibid.* In his 1818 memoir on rotatory polarization, Biot mentioned an unpublished paper of Arago, in which the latter had explained the effect of a quartz slice on polarized light as: ". . . pouvait être considéré comme un faisceau blanc dont les élémens prismatiques auraient été polarisés par des cristaux ayant leurs sections principales dirigeés dans des angles divers." J. B. Biot, *Paris, Mém. Acad. Sci.* 2 (1817): p. 43.

[43] J. B. Biot, *Paris, Mém. de l'Inst.* part II (1812): pp. 264–274.

[44] J. B. Biot, *Paris, Mém. Acad. Sci.*, 2 (1817): pp. 48 ff. In this investigation, he did not use a monochromatic light source but rather a glass filter (e.g. red glass for red light) after the passage of the light through the quartz. *Ibid.*, pp. 53–54.

[45] *Ibid.*, pp. 45–46, 91 ff.

[46] *Ibid.*, p. 125.

[47] *Ibid.*, pp. 131–132.

[48] Herschel had been interested in optical phenomena prior to his study of rotatory polarization. For biographical details of his early life, *cf.* the article by A. M. Clerke in the *Dictionary of National Biography*, ed. L. Stephen and S. Lee (22 v., reprinted, Oxford, Oxford University Press, 1921–1922) 9: pp. 714–719.

quartz was also the only crystalline medium in which rotatory polarization could be produced.[49]

Herschel found that, on certain non-symmetrical prismatic varieties of quartz which Haüy had labeled "plagihedral" and which possessed optical rotatory power, there were secondary faces which, as Herschel put it, "lean or tend, as it were" to the left or to the right of the principal axis. In those crystals where the secondary faces leaned to the left, the plane of polarization was rotated in one direction; conversely when the secondary faces leaned to the right, the rotation was in the opposite direction. Herschel concluded that: "... *these faces are produced by the same cause which determines the displacement of the plane of polarization of a ray traversing the crystal parallel to its axis.*"[50]

He assumed with Biot that the cause of rotatory power resided in the molecules, but he went on to suggest that whatever was responsible for it might, in this case at least, also affect the aggregation of the quartz molecules and therefore, the crystal form of the optically active quartz crystals:

Now so far as the action of crystallized media on light has been examined, there appears to exist an intimate connexion between the crystallographical and optical properties of bodies, and as we have every reason to imagine that the forces by which the particles of matter act on light and on each other, do not differ essentially in their nature, it is easy to conceive that any deviation from perfect symmetry in the distribution of even subordinate forces of any kind, will, in some degree influence the molecules in their mode of aggregation with one another, and however feeble, yet, being a cause constantly in action, may possibly, under favorable circumstances, produce a sensible modification and deviation in their crystal forms. . . .[51]

He suggested that this asymmetry in aggregative force might cause an asymmetry in the molecular decrements and hence, in the crystal form.

Shortly after Herschel's discovery, Augustin Fresnel, the architect of the mathematical wave theory of light, proposed somewhat more specific ideas about the possible internal structure of quartz. Fresnel had investigated rotatory polarization as part of his comprehensive treatment of all optical phenomena. In 1817 he had succeeded in mimicking the optical activity of quartz with a system of glass and crystalline prisms. Subsequently he explained it by means of his undulatory theory of light in terms of the production of two "circular" polarizations, each in opposite directions but which traversed the quartz crystal section with unequal velocities.[52] Unlike Biot, Fresnel was primarily interested in rotatory polarization as explicating the nature of light and the luminiferous ether, rather than for elucidating problems of molecular structure and composition. Yet these two interests were in fact not completely separable, and Fresnel was drawn to derive implications about the molecular arrangement in quartz crystals in order to account for the inequalities in the forward velocities of his two circular polarizations:

The mechanical definition which we have just given of circular polarization enables us to conceive how the peculiar double refraction which rock crystal exhibits in the direction of its axis, can take place: it is that the arrangement of the molecules of this crystal is apparently not the same from right to left as from left to right, so that the light ray whose circular vibrations are executed from right to left, gives rise to an elasticity or propagative force slightly different from that excited by the other ray whose vibrations are from left to right.[53]

Even more specifically on crystal structure:

. . . however, perfectly crystallized substances such as rock crystals present optical phenomena which cannot be reconciled with the complete parallelism of molecular rows and which seem to indicate a progressive and regular deviation of these rows in passage from one section of the medium to the following section.[54]

Curiously enough, Fresnel made no mention of Herschel's discovery, nor did Herschel himself subsequently make the connection between his discovery and Fresnel's ideas on the molecular structure of quartz, even though Herschel became fully acquainted with Fresnel's explanation of rotatory polarization.[55] More-

---

[49] Although he subsequently changed his position, in 1818 Biot believed that rotatory polarization in quartz was a molecular property—as in organic liquids and vapors—rather than a property of the crystalline aggregation. *Paris, Mém. Acad. Sci.* **2** (1817): p. 45.

[50] J. W. F. Herschel, "On the Rotation Impressed by Plates of Rock Crystal on the Planes of Polarization of the Rays of Light, as Connected with Certain Peculiarities in its Crystallization," (dated March 15, 1820; read Apr. 17, 1820), *Cambridge, Phil. Soc. Proc.* **1** (1822): pp. 46–49 (for argument), 49–50 (for quotation).

[51] *Ibid.*, pp. 45–46.

[52] A. Fresnel, "Seconde mémoire sur la double refraction" (presented to the Institut, 22 Jan., 1822), *Œuvres complètes d'Augustin Fresnel*, ed. H. de Senarmont, E. Verdet, L. Fresnel (3 v., Paris, Imprimérie Impérial, 1866–1870) **2**: pp. 504–506; "Mémoirè sur la double refraction que les rayons lumineus éprouvent en traversent les aiguilles de cristal de roche parallelement à l'axe" (presented to the Institut, 9 Dec., 1822), *ibid.* **1**: pp. 748–751.

[53] *Ibid.* **2**: p. 506. Just prior to this passage he had written: "On conçoit que cela [the difference in velocity of the two circular vibrations] peut résulter d'une constitution particulière du milieu réfringent ou de ses molécules intégrantes qui établit une différence entre le sens de droite à gauche et celui de gauche à droite; tel serait, par exemple, un arrangement helicoïdal des molécules du milieu qui offrirait des propriétés inverses selon que ces helices seraient *dextrorsum* ou *sinistrorsum*. *Ibid.*, pp. 505–506.

[54] A. Fresnel, "Extrait du second mémoire sur la double refraction" (published, 1822), *ibid.* **2**: p. 477.

[55] J. W. F. Herschel, "Light," *Encyclopaedia Metropolitana or Universal Dictionary of Knowledge*, ed. E. Smedley, H. J. Rose (29 v., London, B. Fellowes, F. and J. Rivington etc., 1817–1845) **4** (1845): pp. 551 ff.

over, Fresnel was not particularly concerned with speculating in any greater detail about whether his dissymmetry was a molecular or an aggregative property, the issue which was most to exercise Biot over quartz.

Biot himself returned to the subject of rotatory polarization in 1832. In the interval since 1818, although his chief interests had lain elsewhere, he had made observational refinements which enabled him both to observe optical activity under much more refined conditions, and therefore in a far larger number of substances than had been possible up to 1818, and to use this property as a test for change in chemical composition. Regarding Herschel's discovery, Biot acknowledged it and indeed admitted that he had verified it himself. Beyond this, he was rather noncommittal, promising to discuss Herschel's optical-crystallographical correlation more fully subsequently.[56] In fact, already by 1832, Herschel's discovery had pointed up for Biot what seemed to be a peculiarity in the occurrence and nature of rotatory polarization in the inorganic in contrast with the organic kingdoms, for quartz remained both the only inorganic and the only crystalline substance found to possess it.

In 1818 Biot had considered optical activity in quartz to be, as in his organic liquids and solutions, a property of quartz molecules, as opposed to one of molecular aggregation. His reasoning then was related to his contemporary investigations of double refraction. Double-refraction structure seemed to Biot the phenomenon of crystalline aggregation *par excellence*. But rotatory polarization and double refraction were antagonistic to each other, in the sense that the former was only visible along the axis of the crystal where no double refraction occurred. Since Biot reasoned that the absence of rotatory polarization in any other direction in the quartz crystal was due to the inhibitory effect of the double refraction, and since this latter was clearly aggregative in origin, it seemed more likely that rotatory polarization pertained to the quartz molecules themselves, rather than to their aggregation.[57]

By 1832, however, Biot had begun to change his mind on the nature of optical activity in quartz. The fact that on the one hand other types of non-crystalline minerals with the same composition as quartz—opal for example—did not exhibit this property, while an organic substance like cane sugar, both in solution and in its non-crystalline solid state, did show it, indicated to Biot that there might be, after all, something specially relevant about the crystalline state to optical activity in quartz.[58]

Yet, although Biot was tending to see optical activity in quartz in this manner, this did not lead him to embrace Herschel's discovery with the enthusiasm one might have expected. For it was not clear to Biot exactly what this meant, especially since he claimed to possess some rare plagihedral quartz crystals which possessed both left- and right-inclined secondary facets.[59] Biot had considered the possibility of looking for optical activity in crystals of some of his organic liquids and solutions, such as the crystalline form of cane sugar, but he had failed to find it. Although he did not rule out the possibility of its being present in organic crystals,[60] clearly to Biot the significant point was that the rotatory polarization was a property of the individual molecules in optically organic substances and of the aggregative state in quartz.

This distinction between rotatory polarization in quartz and in organic substances was brought out even more sharply by Biot when he finally did get around to publishing his views on quartz, in 1839. The impetus for the publication, interestingly enough, came from an achievement made by Gaudin, who had succeeded in fusing quartz for the first time.[61] The ability to fuse quartz now gave Biot the opportunity to compare the power of rotatory polarization of the same optically active sample, both in crystallized portions where orderly aggregative structure obtained and in fused portions where it presumably had been destroyed. In the fused portions, Biot found no evidence of rotatory polarization, whereas the crystallized portions rotated the plane of polarized light in the usual manner. This, plus the accumulation of other evidence reinforced Biot's conclusion that:

All these facts combine and agree to show that this power in crystals of quartz arises from the habitual mode of superposition of their crystalline laminae and not from the configuration or from the nature of the chemical groups [molecules] which compose them.[62]

---

[56] J. B. Biot, "Mémoire sur la polarisation circulaire et sur ses applications à la chimie organique," *Paris, Mém. Acad. Sci.* **13** (1835): p. 44.

[57] J. B. Biot, *Paris, Mém. Acad. Sci.* **2** (1817): pp. 44–45, also pp. 114–115. He did give expression to one difficulty the molecular view would entail: ". . . enfin, puisqu'il existait des aiguilles à rotations contraires, il fallait qu'elles fussent composées ou au-moins uniformement mêlées de deux substances de nature différente, sans qu'aucun indice dans leur transparence ou leur forme cristalline pût faire soupçonner cette diversité." *Ibid.*, p. 45.

[58] J. B. Biot, *Paris, Mém. Acad. Sci.* **13** (1835): pp. 137–138.
[59] *Ibid.*, p. 44.
[60] Since most organic crystalline substances (e.g., cane sugar) were biaxial, Biot felt that it would be virtually impossible to detect rotatory polarization in these crystals themselves. *Ibid.*, pp. 126–128. Also, cf. "Sur la cause physique qui produit le pouvoir rotatoire dans le quartz cristallisé," *Paris, Comptes rendus* **8** (1839): p. 690.

[61] Cf. M. A. Gaudin, "Formation artificielle des minéraux:—Lentilles en cristal de roche fondu," *ibid.* **5** (1837): pp. 802–803, and "Seconde lettre de M. Gaudin, sur les propriétés du cristal de roche fondu," *ibid.* **8** (1839): pp. 711–713, where he wrote: ". . . je n'ai pu m'assurer s'il y avait ou non déplacement du plan primitive de polarisation. J'envoie, pour décider la question, deux pièces de même epaisseur tirées du même cristal et travaillés ensemble . . . ," *ibid.*, p. 712.

[62] J. B. Biot, *Paris, Comptes rendus* **8** (1839): p. 687. Here

The matter of the correspondence between the asymmetrical crystal faces and the optical activity was a more difficult point for Biot to get around, but the evidence seemed to suggest to him even here that these faces, while no doubt intimately linked with the appearance of optical activity in quartz, were not themselves molecular in origin and, indeed, did not necessarily permeate very deeply into the quartz crystalline structure.[63]

Biot's conviction that rotatory polarization in organic substances was a property of their individual molecules led him to see in this property a unique guide to chemical problems by ascertaining chemical similarities and differences between diverse organic substances and determining whether or not true chemical change had taken place in solutions. He thought that any change in the nature and intensity of optical activity indicated a change in the molecular make up of the substance. Thus in the 1830's Biot attempted to place rotatory polarization at the service of chemistry much as he had earlier tried with double refraction structure.

In particular, he concentrated on solutions of tartaric acid and its compounds. This acid was of interest to Biot for a number of reasons. First, the rotatory dispersive effects of tartaric acid were anomalous. All the other substances which exhibited rotatory polarization followed the rule he had observed for quartz: rotation was greatest for the violet end of the spectrum and least for the red. But tartaric acid in aqueous solution rotated green light the most and violet the least, while different dispersive effects were observed for compounds of tartaric acid.[64] Second, Biot discovered a simple set of linear relations between the concentration of tartaric acid in solution and these dispersive effects.[65]

The change in optical activity of tartaric acid with variations in concentration and in new combinations, seemed to afford a particularly good guide to the chemical situation obtaining in these solutions. In 1838 Biot published two long studies on the optical activity of tartaric acid and its combinations, the results of two years of investigation. They were directed at elucidating what were old questions for Biot about the nature of chemical combination: did it take place (in solution) in definite proportions? And when an acid was put in the presence of two bases in solution, did it combine only with one of them, or was it shared in varying proportions by them both?[66] These were the old Bertholletian issues and Biot himself thought that he had found once again proof in optical phenomena for his Berthollietian beliefs.

His characterization of chemical reaction in solution echoed Berthollet in the use of a celestial "planetary" analogy, interestingly enough also used by Dumas and Laurent at about this time in their first formulations of the type theory:

The experiments which I am going to relate prove, I think, clearly that, in liquid systems, combination does not at all operate by discontinuity; and that all the particles which are brought together react simultaneously one equally on the other, as do the celestial nebulae whose constituent planets interact together without ever coming into contact.[67]

As a general conclusion, Biot claimed that the continuity of change in rotatory power which he had observed, affording as it did unique insight into molecular constitution, proved the existence of varying proportions

---

Biot summarized all the evidence which he considered as countering the assumption that quartz's optical activity was molecular: Fresnel's successful imitation of it with glass plates (p. 684) the existence of non-crystalline quartz minerals, such as opal, which did not exhibit optical activity (p. 685), the lack of evidence for the existence of two acids of silicon (corresponding to the two directions of optical activity of quartz [p. 686], this latter prescient of Pasteur), and the existence of optical activity in opposite directions in different regions of the same mineral sample (p. 687).

[63] *Ibid.*, pp. 688, 689. Herschel himself seems to have, in part, misunderstood Biot's arguments, when he wrote: "From these circumstances [optical activity in organic liquids] he [Biot] concludes that the rotatory power is essentially inherent in the molecules of bodies and carried with them into all their combinations. But this is too rapid a generalization; for neither sugar nor camphor in the solid state possess this property, though examined for it in the same circumstances as quartz is, by transmitting the polarized ray along their optic axes; and, on the other hand, quartz held in solution by potash or (as Dr. Brewster has found) melted by heat, and thus deprived of its crystalline arrangement, manifests no such property. This obscure part of chemical optics well deserves additional attention." J. W. F. Herschel, *loc. cit.*

[64] J. B. Biot, "Mémoire sur plusieurs points fondamentaux de mécanique chimique," *Paris, Mém. Acad. Sci.* 16 (1838): pp. 254–256.

[65] The amount of rotation for each color varied with the nature and proportion of the solvent and with the temperature. For aqueous solutions, the relation Biot established was:

$$\frac{\alpha}{l\epsilon\delta} = A + Be$$

where:

α = deviation of the ray
l = length of tube through which one observed the polarized light
δ = actual density of the solution
ϵ = weight proportion of acid
e = weight proportion of water

$$[e + \epsilon = 1]$$

A and B are constants which vary for each color. For a given color, A changes with temperature. J. B. Biot, Méthodes mathématiques et expérimentales pour discerner les mélanges et les combinaisons chimiques définies ou non définies, qui agissent sur la lumière polarisée" [presented 11 Jan., 1836], *Paris, Mém. Acad. Sci.* 15 (1838): pp. 229–231. *Cf.* also *ibid.* 16 (1838): pp. 257–258.

[66] *Ibid.*, pp. 229–230. *Cf.* also, *ibid.* 15 (1838): pp. 93–279.
[67] J. B. Biot, *Paris, Mém. Acad. Sci.* 16 (1838): pp. 261–262.

in the compounds tartaric acid formed in solution:

> The experiments contained in this memoir seem to me to resolve neatly that so-controversial question of combination in definite or indefinite proportions, by reducing it to the following proposition.
>
> When tartaric acid, water and alkalis, or tartaric acid, water and alumina are brought together *in a fluid state*, either under certain conditions or in certain proportions, these three substances unite immediately and form a uniform molecular system endowed with special physical properties which depend on their existent proportions and vary *continuously* with these proportions. Combinations which exhibit these characteristics are thus not restricted to conditions of discontinuity which are observed in solid and crystalline salts, which separate out from the same medium. And, on the contrary, these molecular groups constituting the salts decompose and lose their fixity of proportions, when they are returned to the fluid state.[68]

Another issue confronting chemists with increasing persistence in the 1830's and to which Biot attempted to bring clarification through his rotatory polarization studies was isomerism. Here also tartaric acid was in the forefront. Indeed, this acid and its isomer, racemic or paratartaric acid, became the *locus classicus* of isomerism.

The earliest recognized substances with the same chemical composition but different chemical properties were found in the middle of the second decade of the century. During the 1820's, the number of these substances proliferated, the most spectacular case appearing to be the urea-ammonium cyanate identity established by Wöhler. It was a moot point whether isomerism and polymorphism were both subspecies of the same phenomenon.[69] In 1830 Berzelius gave a general survey of these groups of substances in the paper in which he coined the term "isomeric" meaning "composed of equal parts," to denominate them.[70] The focus of this paper was on tartaric and racemic acid and some of their salts.

In the early 1830's even more cases of what were thought to be isomers among organic compounds were found and Biot turned his attention to this subject. He and a young colleague, the chemist Jean François Persoz, subjected a large number of these substances to the test of optical rotatory power in order to determine decisively their similarities and their differences. Biot and Persoz also studied the conversion of starch into sugar in the maturation of fruits, using the change in nature and degree of optical rotatory power of each successive product as an indication of chemical change. In the course of these studies, they isolated an intermediate product which appeared in the acid solutions immediately following the breakdown of the starch globules which they named "dextrin" because of its exceptional rotatory polarization effect to the right.[71]

In the case of tartaric and racemic acid, the former exhibited strong rotatory power whereas the latter was optically inactive. In his studies on tartaric acid, Biot himself was relatively unconcerned with following the comparison with its isomer further. But these were two organic substances which happen also to crystallize and, therefore, whose crystal forms could be correlated with their optical rotatory properties even if optical activity could not be observed directly in the crystals.[72]

Back in 1830, when Berzelius was studying the isomerism of the tartrates and racemates, he had recognized that the differences between them, if due to some difference in the arrangement of the constituent atoms, ought to be manifest in their crystal forms as well as in their properties. He had then written to Mitscherlich, asking him to test this with crystals of some neutral salts of racemic and tartaric acid, the acids themselves being in Berzelius's opinion less useful because of their different amounts of water of crystallization. Berzelius wrote that he himself had crystallized the double sodium and potassium salts, but he was too deficient in crystallographic knowledge to test them. If the forms of these neutral salts differed, "then the hard nut of the dual relationships of bodies of the same composition would be cracked in a simple and perhaps correct manner." [73]

---

[68] *Ibid.*, p. 385. Louis Pasteur took up these Bertholletian issues in his physics thesis. *Cf.* below, chap. 6.

[69] *Cf.* J. R. Partington, *op. cit.* 4: pp. 256–259. For the structural significance of the synthesis of urea, *cf.* J. H. Brooke, "Wöhler's Urea, and its Vital Force?—A Verdict From the Chemists," *Ambix* 15 (1968): pp. 108 ff. J. B. Biot, *Paris, Mém. Acad. Sci.* 13 (1835): pp. 141–142.

[70] J. J. Berzelius, "Composition de l'acide tartrique et de l'acide racémique (traubensäure)," *Annal. de chimie* 46 (1831): pp. 113–147 (p. 136 for term, "isomeric").

[71] J. B. Biot, *Paris, Mém. Acad. Sci.* 13 (1835): pp. 139–169; J. B. Biot and J. Persoz, "Mémoire sur les modifications que la fécule et la gomme subissent sous l'influence des acides," *ibid.*, pp. 437–496 (Read 14 Jan., 1833). Persoz and J. Payen went on to discover the enzyme diatase.

[72] In another paper dealing specifically with the use of rotatory polarization to study isomerism, Biot wrote of tartaric and racemic acid: "Néamoins, leur mode différent de cristallisation, sourtout leur solubilité inégale quand ils sont désagregés, inégalité que se communique à plusieurs de leurs sels, suffit pour prouver indubitablement que leur constitution moléculaire est différente; et ce sont là, je crois, les seuls caractères de dissemblance par lesquels la chimie les distingue." "Sur l'emploi de la lumière polarisée pour manifester les différences des combinaisons isomériques," *Annal. de chimie* 69 (1838): p. 27. He then immediately referred to their difference in optical activity.

[73] In the German translation: "Ist die Form verschieden, so ist der schwere Knoten von dem doppelten Verhältnissen der gleich zusammen desetzten Körper in einer einfachen und vielleicht richtigen Weise gelöst...." Letter of Berzelius to Mitscherlich, Stockholm, May 28, 1830, E. Mitscherlich, *Gesammelte Schriften*, p. 92. *Cf.* also: J. D. Bernal, *op. cit.*, p. 189 (his translation): "Should the form be different, then the awkward difficulty (*der schwere Knoten*) of the dual relationship of bodies with the same composition, would be resolved in a simple and possibly correct manner."

Within a month, Mitscherlich wrote back to say that the crystallographical investigations were underway and had so far shown only differences between the forms of the corresponding salts. He was particularly eager to pursue this investigation because of what it shows about "the form of their combination."[74] Berzelius himself discontinued this study soon after,[75] and Mitscherlich did not produce anything until the following summer of 1831. He had not found any cases of crystallographic identity between corresponding tartaric and racemic salts, except one: the sodium-ammonium double salts of each. In line with his interest in elucidating chemical binding by these studies, Mitscherlich had hoped to use these salts to convert racemic into tartaric acid, but had been unsuccessful.[76] Berzelius, recognizing the anomaly that this posed for his concept of isomerism, urged Mitscherlich to continue his studies of this subject.[77]

The discovery of the apparent crystallographic identity of the sodium-ammonium double tartrate and racemate salts occurred after Berzelius's paper had been published, and Mitscherlich neither saw fit to publish it himself immediately, nor, apparently, to continue with this line of investigation. For over ten years, he was silent on the subject.

It was in this interval that Biot had returned to his study of optical activity and carried out his great series of researches of the 1830's. It must be borne in mind that when Berzelius and Mitscherlich dropped the problem of the crystal form of tartrate-racemate salts in 1831 Biot had not yet studied their optical activity. It was apparently Biot's investigation into optical activity generally, and its use in scanning the course of chemical reactions in solutions particularly, that brought Mitscherlich back to the question of tartrates and racemates in the 1840's, this time taking into account the additional factor of the differences in their optical rotatory powers.

In 1843 he published a short paper on some chemical investigations he had been carrying out on fermentation—a subject which was to be of great significance to Pasteur—and using the optical activity of three sugars in much the same way Biot and Persoz had done earlier in their studies on the conversion of starch into sugar in fruit. Near the beginning of this paper Mitscherlich discussed the optical activity of tartaric acid and its changes with combination which he took, as had Biot previously, as a paradigmatic compound for this property. But his point of view was somewhat different from Biot's, as one might expect from Mitscherlich's crystallographical orientation:

The property of optical activity belongs only to tartaric acid, and it stems from the grouping of the atoms, just as one is led to conclude from the relationship of the crystal form of quartz to its optical activity. Racemic acid, citric acid, succinic acid and many other acids neither possess the property themselves or in combinations.[78]

Implicit in this statement was the problem Mitscherlich considered twelve years before: did any of the tartrate salts have the same crystal form as the corresponding racemate salts, even though the one exhibited optical activity and the other did not? The following year, Mitscherlich communicated to Biot a note summarizing his investigations—or reinvestigations—of just this question with regard to the corresponding sodium-ammonium double salts. The results reiterated and reinforced the dilemma of 1831, for Mitscherlich had been able to find no differences in the chemical composition, specific weights, optical structure, or crystal form of the tartrate and racemate. He noted Biot's discovery of their difference in optical activity, even though, as Mitscherlich put it, the other, particularly the crystallographic characteristics, indicated that ". . . the nature, and the number of atoms, their arrangement and their distances, are the same in the two substances under comparison."[79] Biot presented Mitscherlich's note to the Académie des Sciences and offered his own comments. He took the opportunity to reassert more strongly than ever his belief that optical rotatory activity took priority over all other physical and chemical characteristics for getting at the true nature of the constituent molecules:

The only phenomenon whose observations and measure could legitimately be related to the constituent molecular groups themselves consists *uniquely* in the deviation impressed upon the polarization planes of light rays, independently of their fortuitous state of aggregation, by a great number of substances, in truth, of organic origin.[80]

It was, to Biot, invalid to use other physical and chemical properties like those cited by Mitscherlich in order to deduce anything about the individual molecules, since

---

[74] Letter of Mitscherlich to Berzelius, Berlin, June, 1830. E. Mitscherlich, *Gesammelte Schriften*, p. 93.

[75] Letter of Berzelius to Mitscherlich, Stockholm, July 13, 1830. *Ibid.*, p. 94.

[76] Letter of Mitscherlich to Berzelius, Berlin, June 16, 1831. *Ibid.*, p. 96.

[77] Letter of Berzelius to Mitscherlich, Stockholm, June 28, 1831. *Ibid.*, p. 97.

[78] E. Mitscherlich, "Ueber die Gährung," *ibid.*, p. 537. Spurred on by Mitscherlich's investigations of optical activity, Biot published a very lengthy summary of his ideas and his investigations since the late 1830's of chemical reaction, especially of tartaric acid, in solution. "Sur l'emploi de la lumière polarisée pour étudier diverses questions de mécanique chimique," *Annal. de chimie* 10 (1844): pp. 5–53, 307–327, 385–402; 11 (1844): pp. 82–112. *Cf. ibid.* 10: p. 5 for his introductory reference to Mitscherlich. In *ibid.* 11: p. 108, he wrote: "Si je ne me fais pas illusion sur la portée de ces phénomènes, c'est là un champ d'études des plus importants, comme des plus fructueux, dans lesquels un chimist physicien puisse entrer."

[79] J. B. Biot, "Communication d'une note de M. Mitscherlich," *Paris, Comptes rendus* 19 (1844): p. 720.

[80] *Ibid.*, p. 722.

these properties depended upon molecular aggregation as well as molecular nature and it was impossible to separate out effects of the individual molecules from those of their aggregate for any other phenomenon than optical rotatory power.

Biot was, here, reasoning in a kind of inverse fashion from his views on optical activity in quartz: there, he had proved to his own satisfaction that both the asymmetrical crystal form and the optical rotatory activity were properties of molecular aggregation. In the case of tartaric acid (and the other organic optically active substances) where the property was one of the individual molecules, there was no necessity that crystal form, optical structure, and the other physical and chemical properties had to correlate with the optical rotatory activity.

No necessity, perhaps, but even Biot recognized that the difference in molecular nature as revealed by the optical activity of their solutions, ought to have *some* expression in the aggregative characteristics of the crystals. He ended his comments to Mitscherlich's note with words which were to be more prescient than he could then have realized:

But it would be difficult to conceive mechanically how *dissimilar* molecules, placed in the same number and at equal distances, and arranged in the same manner, could produce material systems of similar types, whose crystal form and physical properties are as exactly parallel as the two substances under comparison here; at least nothing gives assurance of it and the contrary would be much more presumable.[81]

Within four years, Biot's surmise was to be proved by Pasteur.

VI. PASTEUR'S DISCOVERY OF ENANTIOMORPHISM: THE SYNTHESIS OF A TRADITION

In 1841 Frederick Hervé de la Provostaye published the first of several papers dealing with various crystallographical studies he had made. This one dealt with the crystal forms of tartaric and racemic acid, as well as those of some of their salts. This was the first comprehensive survey of the crystallography of these compounds to be published.[1]

We have already had occasion to mention De la Provostaye in connection with his forestalling Laurent in publishing a study of isomorphism between crystals of substitution compounds in 1840.[2] His investigation of tartrate and racemate crystals was an extension of the work on isomorphism and an extension along lines parallel to those pursued by Laurent. The similarity of De la Provostaye's orientation and motivation in his new research to that of the proponents of the crystallographical-chemical approach discussed in chapter IV was made explicit by him in his opening comments:

In the actual state of science, everything leads one to think that physical and chemical properties of substances are determined much less by their individual nature than by the relative disposition of the molecules, their mode of grouping and aggregation. But one realizes that there is another effect, deriving from the same source in a manner more direct still; this is the external form of a substance when it is obtained pure and crystallized.[3]

De la Provostaye was using tartrate and racemate crystals to treat such questions as: Were crystals of different tartrates isomorphic with one another? What could one conclude about the isomorphism of the various atomic complexes which combined with the nuclear tartrate radical? Were *isomeric* tartrates and racemates also isomorphic?[4]

To answer these questions, he examined carefully the crystal forms of some tartrate and racemate compounds. His results were disappointing. He found no consistent isomorphism either between crystals of the different tartrates or between tartrates or racemates. Some of each crystallized as right prisms, others as oblique prisms, incompatible crystal forms in the prevailing crystallography.

He did note in passing that the crystals of the emetic of ammonia (double tartrate of antimony and ammonia) and tartaric acid were hemihedral, but he made nothing of this oddity.[5] Despite his unspectacular results, De la Provostaye had concluded his introductory remarks to this study with optimistic and ironically prophetic words:

Chemists will discover, in carefully made crystals a useful character, a kind of definition of the substance which they study; their theories will be able to find here a point of appeal and a means of control. Finally, the formal analogy will give birth in their minds to new relations, to points of view unperceived until now.[6]

In spring, 1848, while carrying out researches in Antoine Jerome Balard's laboratory at the École Normale, Louis Pasteur made two major discoveries on the same day which were to catapult him to instant scientific renown. Pasteur had been studying the crystal forms of various tartrates and racemates, using De la Provostaye's measurements as his guide. But Pasteur was also dealing with an issue with which De la Provostaye had been unconcerned, but which Mitscherlich had brought to a head in his note to Biot of 1844: the

---

[81] *Ibid.*, p. 723.
[1] F. Hervé de la Provostaye. *Annal. de chimie* 3 (1841): pp. 129–150. Dumas had given him the crystal samples. *Ibid.*, p. 150, note.
[2] *Cf.* above, chap. 4.

[3] F. Hervé de la Provostaye, *Annal. de chimie* 3 (1841): p. 129.
[4] *Ibid.*, pp. 149–150.
[5] *Ibid.*, p. 132 and pl. I, fig. 2; pp. 145–146 where he spoke of "hémiédrie."
[6] *Ibid.*, p. 130.

physical and crystallographical identities of sodium-ammonium tartrate and racemate versus the difference in their optical activity.

Pasteur's first discovery was something that Mitscherlich and Biot in turn had apparently overlooked, and De la Provostaye had not been in a position to appreciate. Tartrate crystals generally, and the sodium-ammonium tartrate particularly, had small asymmetrically placed facets on some of their edges. They were hemihedral in the same way as plagihedral quartz crystals, but, unlike quartz, all the tartrate crystals were hemihedral in the same sense, just as they all rotated the plane of polarized light in the same direction.

Turning to the corresponding racemate, Pasteur had expected to find symmetry to match their lack of rotatory power. But, to his surprise and puzzlement, he found that his racemate crystals, too, were hemihedral. However, he made another discovery: he noticed that the hemihedry in the racemates was not all in the same sense. Some had secondary facets orientated in the same sense as the tartrates. These racemates were formally indistinguishable from the corresponding tartrates, and some had these facets orientated in the opposite sense. The formal relationship of these two types of racemate crystals, like that of the two hemihedral forms of quartz, was the same as a left and right hand. They were non-superposable mirror images of each other.

Pasteur carefully separated the two forms of racemate crystals and dissolved each form separately to test for optical activity. The solution of racemate crystals whose hemihedry was in the same sense as the tartrates deviated the plane of polarized light in the same direction and virtually the same amount as the solution of the corresponding tartrates did. The solution of the oppositely orientated hemihedral racemates was optically active in the other direction.[7]

With these discoveries, the puzzle posed by Mitscherlich in 1844 was essentially solved, the last short step being the demonstration that the two forms of racemate crystals, when dissolved together, neutralized each other's rotatory power to give the characteristic optical inactivity of racemate solutions.

Pasteur's discovery of enantiomorphism in racemate crystals—nonsuperposable mirror images—and his correlation of these crystal forms with optical activity, was the first great achievement of the man who, perhaps more than any other, is the exemplar in the popular mind of the successful experimental scientist. And this first set of discoveries is itself something of a paradigm of the experimental method. But, like all great creative achievements in science, the path towards Pasteur's discovery of enantiomorphism was much more complex than first appearances would lead one to believe. Much more was involved than merely noticing some additional details about tartrate crystal forms that De la Provostaye, Mitscherlich, and Biot had overlooked. As Pasteur himself later[8] said, the mind had to be "prepared" for discovery. In Pasteur's own case of enantiomorphism, the preparation seems to have lain in the particular orientation of his scientific interest in the period just preceding his discovery—an orientation towards the elucidation of molecular structure through the combined use of crystallography and chemistry much along the lines of Delafosse and Laurent. What we shall now proceed to do is to reconstruct, insofar as documentary evidence permits, the background to Pasteur's great discovery, paying particular attention to the play over him of the influences of Delafosse and Laurent, with each of whom Pasteur came successively into contact during the course of his scientific training.

Pasteur's career in science commenced in earnest with his coming up to Paris as a student at the École Normale in 1843. Here, he received a comprehensive grounding in the physical sciences, attending among other courses Dumas's lectures in chemistry and Delafosse's in mineralogy. He obtained his baccalaureat in 1846 and was then taken on as an *agregé* by Balard, the professor of chemistry at the École Normale. Pasteur remained with Balard until the fall of 1848. In the summer of 1847, he completed and wrote his thesis for the doctorate. In the following spring, he made his discovery of enantiomorphism and its correlation with optical activity.[9]

As to the origins of his interest in the question of crystal form and optical activity, Pasteur himself asserted that it went back to his student days at the École Normale, when by chance he came upon Mitscherlich's note of 1844 which instantly both fascinated him and puzzled him:

I meditated for a long time on that note; it troubled all my student ideas; I could not understand how two substances could be so similar as Mitscherlich said without being completely identical. To be capable of surprise is the first movement of the spirit towards discovery.[10]

---

[7] *Cf.* J. D. Bernal's account in *op. cit.*, pp. 195–206.

[8] The actual context of Pasteur's "prepared" quotation referred to the relation of pure to applied science. L. Pasteur, "Pourquoi la France n'a pas trouvé d'hommes superieurs au moment du peril," [La salut public, Lyon, mars, 1871]. *Œuvres de Pasteur*, ed. P. Vallery-Radot (7 v., Paris, Masson, et 6¹ᵉ, 1922–1939) 7: p. 215. The quotation has entered the popular domain misconstrued as a statement of scientific methodology, and it is in its popular sense that I refer to it. Pasteur used the expression, "idée préconçue" for hypotheses which served as guides for research. *Cf.* L. Pasteur, "Recherches sur la dissymétrie molécularie des produits organiques naturels (Leçons professées à la Société chimique de Paris le 20 janvier et le 3 fevrier 1860)," *Œuvres de Pasteur* 1: p. 322.

[9] Gerald L. Geison has provided a good biographical sketch of Pasteur's career in his article on Pasteur in the *Dictionary of Scientific Biography* 10 (1974): pp. 350–416.

[10] L. Pasteur, "La dissymétrie moléculaire" (Conférence faite à la Société chimique de Paris le 22 Decembre 1883), *Œuvres de Pasteur* 1: p. 370.

Pasteur gave no more details of exactly when in his studies at the École Normale this epiphany took place and there seems to be no additional surviving evidence to help pinpoint the stage he had reached in his scientific education at this time.

Yet there is evidence that the source of presentation of this issue in his student courses at the École Normale was the mineralogy course given by Delafosse. As we have already seen, the early 1840's was the time when Delafosse was elaborating his own speculations about atomic arrangements in crystals, marked especially by his detailed paper of 1843 focused on the subject of hemihedry.

It will be recalled that the main point of this paper was to refine Haüy's molecular theory of crystal structure, one of his refinements being his recognition that the external form of the crystal must reflect the form of the true physical crystalline molecule. With regard to the particular subject of this paper, hemihedral crystals must be composed of hemihedral molecules, especially since hemihedry was also associated with closely related physical characteristics such as pyroelectricity.

Among the hemihedral substances with which Delafosse dealt in this paper was quartz with its attendant optical activity. Recalling Herschel's discovery of the correlation between plagihedral quartz crystals with asymmetrically inclined secondary facets, and their optical activity, Delafosse attempted to provide an explanation for this in line with his general prescription, in terms of the form of quartz molecules. He knew that Biot had distinguished optical rotatory power, dependent on its aggregative state, from that of his organic liquids, vapors, and solutions. But, with his grounding in the molecular theory of crystal structure and with his interest in working out molecular structures, Delafosse refused to accept Biot's distinction. Quartz too was optically active because of the form of its molecules, which was manifested in the form of the crystal.[11]

As an initial assumption about the molecular form of quartz, Delafosse considered it to be rhombohedral like that of quartz crystals in their primitive form. But to account for the asymmetry of the secondary facets and the optical rotatory power, he suggested that there was asymmetry within these molecules—either the atoms themselves were helically arranged in the molecule, or the atoms located at the apices of the lateral solid angles of the molecular rhombohedron were somehow inclined to one side or another, or the rhombohedronal molecules had somehow suffered distortion through a lateral twist of one half with respect to the other.[12]

In his mineralogy lectures of this period, which Pastuer attended, and a set of notes which survive in Pasteur's hand, Delafosse gave considerable attention to his preoccupation with hemihedry. He stressed his own view that this was fundamentally a manifestation of molecular form and he presented some of his own molecular models, like the one for pyrite which we examined in an earlier chapter. With regard to hemihedry in quartz, Delafosse discussed Herschel's discovery and once again defended his own position—specifically against Biot—that Herschel's correlation was an expression of quartz's molecular structure.[13]

Delafosse's crystallographical approach to quartz was a remarkable prevision of Pasteur's own explanation of optical activity and enantiomorphism in the tartrates and racemates. And indeed, Pasteur acknowledged the influence of Delafosse on him in leading him towards the investigation of hemihedry and optical activity:

Guided thus, on the one hand, by the fact of the existence of molecular rotatory polarization, discovered by M. Biot in tartaric acid and in all its combinations, on the other, by the ingenious rapprochement of Herschel, and in the third place, by the knowledgeable views of M. Delafosse, for whom hemihedry has always been a law of structure and not an accident of crystallization, I presumed that there could be a correlation between the hemihedry of the tartrates and their property of deviating the plane of polarized light.[14]

Thus, when Pasteur was graduated from the École Normale in 1846, he quite possibly knew of and had pondered over Mitscherlich's note on the sodium-ammonium tartrate and racemate. He certainly had been exposed to the issue of hemihedry and optical activity in quartz, to Delafosse's general speculations about molecular form and atomic arrangement, and to his par-

---

[11] G. Delafosse, *Paris, Mém. savans étrang.* 8 (1843): pp. 683 ff. Delafosse also rejected Fresnel's ingenious suggestion of a helical aggregation of what Delafosse took to be symmetrical molecules, and this for an interesting reason. It was completely at odds with the crucial point of Haüy's (and Delafosse's) theories of crystal structure: the parallel orientation of all the molecules which compose a crystal. *Ibid.*, p. 685. It was only with L. Sohncke that such helical structures were incorporated into the theory of crystal lattice.

[12] *Ibid.*, pp. 685–688.

[13] Bibliothèque Nationale, Pasteur MSS. "1846. Notes prisés par Pasteur au cours de Delafosse," pp. 31–33 for hemihedry generally, also pp. 27 ff.

In the front of the notebook are seven "Questions générales (sujets de leçons ou d'argumentation)" of which the following are of interest: "2. Rapports de la forme et des propriétés optiques [physiques en général]. 3. Rapports de la forme et de la composition chimique—(isomorphisme—isomerie—polymorphisme.) 6. Quelle influence a exercée sur la classification minéralogique la découverte des faits relatifs à l'isomorphisme, à l'isomerie et au dimorphisme? 7. L'hemiédrie est-elle un fait accidental, et sans importance pour la distinction des systèmes cristallins, et par suite pour celle des espèces minérals?"

[14] L. Pasteur, "Recherches sur la dissymétrie moléculaire" *Œuvres de Pasteur* 1: p. 322. *Cf.* also his reference to Delafosse in his first elaboration of the discovery of enantiomorphism. "Recherches sur les relations qui peuvent exister entre la forme cristalline, la composition chimique et le sens de la polarisation rotatoire" (*Annal. de chimie* 24 [1848]: pp. 442–459), *ibid.*, p. 66; and the uncompleted "Études de chimie moléculaire ou recherches sur la dissymétrie dans les produits organiques naturels," (1878), *ibid.*, p. 398.

ticular views on quartz. But this did not lead Pasteur on to the issue of hemihedry and optical activity in the tartrates with quite the directness or immediacy that his retrospective account suggests. Before he confronted the tartrates, he passed through the influence of Auguste Laurent.

We have already mentioned that Pasteur was taken on as *agregé* to Balard after receiving his degree from the École Normale, where he remained until the fall of 1848. Early in this next period, from 1846 until April, 1847, Laurent, too, worked in Balard's laboratory. Laurent had taken a professorship in chemistry at the University of Bordeaux in 1839, but had thrown it over in 1845 in order to seek a more worthy post in Paris. In the years between his return to Paris and his appointment as assayer at the Paris mint in 1848, Laurent lived very much a hand-to-mouth existence, accepting offers of help from fellow chemists like Balard until he would become dissatisfied with his subordinate position and move on.[15]

The brief contact between Laurent and Pasteur affected Pasteur deeply at the time, as he revealed in his enthusiasms to his friend and confidant, Chappuis:

There is not, I think, any chemist with whom I could learn more, so willingly does he give of his advice. I think, moreover, that M. Laurent is destined to occupy the premier position among chemists in a few years.[16]

Before Laurent left the laboratory of Balard, he engaged Pasteur in research which was destined to become part of Pasteur's chemistry thesis. In subsequent researches and publications—his physics thesis of the summer of 1847, his studies on dimorphism completed early in 1848 —Laurent's influence was strongly evident, most notably through his own contemporaneous concern with isomorphism, polymorphism, isomerism, and their interrelations. It would, I think, not be going too far to suggest that the orientation which Pasteur derived from Laurent was decisive in leading him on both to raise and to solve the problem of the tartrates and the paratartrates.

In the spring of 1847, just before he left, Laurent put Pasteur onto a project dealing with the saturation capacity of arsenious acid and the arsenites of potassium, sodium, and ammonia.[17] With Laurent's departure, this project proved abortive, but Pasteur used the conclusions for the first part of his chemistry thesis, submitted along with his physics thesis for his doctorate in August, 1847. At the conclusion of this first section, Pasteur gracefully conceded to Laurent his dominant role in this project.[18]

But there was a second section of the thesis which for our purposes, is of greater interest than the first. This latter part appears to have been Pasteur's own, but, unlike the first section, which was purely chemical, this one manifested the sort of crystallographical-chemical considerations so characteristic of Laurent. What Pasteur was concerned to do in the final portion was to find corroboration for the chemical conclusions of the first part, in the study of crystal form. His conclusion to the first section was that there existed two types of arsenious acid: a monobasic and a dibasic type. In the second part, he attempted to show that there existed also dimorphic forms of this acid, corresponding to its apparent double chemical state. He thought he had discovered two such varieties which, moreover, had corresponding isomorphic forms in antimonious acid.[19]

---

[15] The details of Laurent's life in this period, and the separation of fact from accusation, etc., remain to be elucidated. But *cf.* S. C. Kapoor, *Dictionary of Scientific Biography* 8: pp. 54–55). Pasteur himself gave a brief and probably objective account of the kind of motivation impelling Laurent to move about, when he wrote to his friend Chappuis, in March, 1847: "Je serais encore occupé à travailler avec M. Laurent, si M. Laurent n'avait eu le projet de quitter l'École en me laissant continuer seul se travail [*cf.* below, footnote 17]. Chimiste indifatigable, il trouvait qu'il perdait beaucoup de temps à venir à l'École le matin, à s'en retournir le soir, et il a résolu de se construire un laboratoire chez lui-même. Cependant, d'après le conseil de M. Balard, peut-être restera-t-il à l'École, ce que je souhaite de grand cœur." Letter to Chappuis, Paris, March 20, 1847, *Correspondance de Pasteur, 1840–1895* (4 v., Paris, Flammarion, 1940–1951) 1: p. 152.

[16] *Ibid.* In this same letter, Pasteur went on to write: "Il [Laurent] vient d'être nommé suppléant de M. Dumas à la Sorbonne. Dans ses deux premières leçons, il a été aussi hardi que dans ses mémoires, et parmi les chimistes ces leçons ont fait beaucoup de bruit. J'ai l'honneur de passer chez M. Dumas les soirées de jeudi, et jeudi dernier ces leçons faisaient le sujet de plusieurs conversations." *Ibid.*
I have recently discovered manuscript notes, in Pasteur's hand, of what are almost certainly these lectures of Laurent. They are in the back of a notebook of Pasteur, the greater part of which is given over to his notes of Pélouze's lectures. Wellcome MS 3774. Pasteur, Louis. "Notes on Chemistry Taken Down from the Lectures of Professor Théophile Pélouze [1807–1867] given at the Collège de France.

[17] Lettre 65: Laurent à Gerhardt, Paris, Feb. 23, 1847, where he wrote: "Il y a un jeune homme [Pasteur] à l'École normale auquel j'ai fait examiner le chlorure d' arsenic ammoniacal." M. Tiffeneau (ed.), *op. cit.*, p. 228.

[18] L. Pasteur, "Thèse de chimie: Recherches sur la capacité de saturation de l'acide arsénieux. Étude des arsenites de potasse, de soude et d'ammoniaque," *Œuvres de Pasteur*, 1: p. 8. *Cf.* also, Letter to Chappuis, Paris, May 12, 1847, *Correspondance* 1: p. 153.

[19] Pasteur made his debt to Laurent's ideas explicit in the introduction to the second part of the thesis: "Je vais établir que l'acide arsénieux et l'acide antimonieux sont à la fois dimorphes et isomorphes. En voyant un acide anhydre sous deux formes très distinctes, on sera plus enclin à croire que de même qu'à l'état libre il s'offre à deux états distincts, il pourrait bien en être de même dans ses combinaisons. Cette idée deviendra une induction très raisonnable dès que M. Laurent aura publié les faits extrêmement remarquables qu'il vient de découvrir dans l'etude de l'acide tungstique." "Thèse de chimie," *Œuvres de Pasteur* 1: p. 15. Laurent tried to establish six "types" among the tungstates, each of whose nuclear anhydrous acid would be composed respectively of one, two ... six molecules, the nuclear formulas being WO^3, W^2O^6 etc. *Cf.* above, chap. 4, footnotes 101–102.

Pasteur's next work was his physics thesis, begun some time in the late spring of 1847 and finished by August, when it was submitted with the chemistry thesis. The completed physics thesis seems to have been something of an afterthought, at least from indications in his letters to Chappuis. Pasteur had originally decided to write something on atomic volumes for this thesis [20] but in July he had changed course, as he wrote in a guarded yet portentous comment to Chappuis:

I shall do little in physics. I shall only set up a program of highly useful research which I shall undertake next year and which I shall only commence in the thesis.[21]

The change was that Pasteur had turned to the study of optical activity. From his student notes, we know that he had been exposed to this subject in his course on optics and of course, to the relation between hemihedry and optical activity in Delafosse's lectures. Pasteur himself offered a unit on optical activity and "saccarimetry" in the lectures he delivered as Balard's *agregé-preparateur* in 1846–1847. So there is no question about his general background and interest in this subject.[22]

Moreover, Laurent had been using polarimetry in his chemical researches while working with Balard. Laurent's interest in optical activity went back a number of years and he had been in contact with Biot over it. It was Biot who, early in the 1840's, had interested Laurent in rotatory polarization as an analytical tool and had given Laurent the use of his own polarimeter to investigate optical activity in two new alkalis which Laurent had synthesized. Although they turned out to be inactive, Laurent had gone on to make some interesting observations. Noting that no artificially prepared organic alkalis had been found to possess optical activity, he carried out a series of experiments to test whether some chemical constituent in the optically active bases, such as oxygen, was responsible for the activity, or whether the natural versus the artificial origins of the alkalis in question had anything do with it.

Testing nicotine, naturally occurring, and aniline, artificially prepared, Laurent found that even though their chemical compositions were very similar, nicotine had a very energetic activity whereas aniline exhibited none. He drew no conclusions, but rather suggested that further investigations be made into the synthesis of artificial nicotine.[23] At the February 15, 1847 *Séance* of the Académie des Sciences, Laurent read a paper in which he made use of optical activity in support of his nuclear theory, still then under attack from Berzelius. He examined the optical activity of various chlorinated and brominated derivatives of cinchonine and demonstrated that there was but slight variation in the quality of their rotations, thus indicating similarity in their molecular makeup.[24]

We do not know what direct influence, if any, Laurent had in turning Pasteur to the investigation of optical activity. But we do know that Pasteur valued highly the significance of Laurent's recent use of it in defense of his nuclear theory for the elucidation of atomic arrangement. This Pasteur made clear in what was something of an introductory manifesto in his physics thesis on the subject of the use of physical and crystallographical data to supplement the chemical information in working out nuclear constitution:

How fecund have the considerations of isomorphism been! I believe that (considerations) of atomic volumes and optical properties will not be less so. M. Berzelius has admitted without question that chlorinated strychnine was analogous to strychnine; that the molecular group was the same in these two substances, from the moment that M. Laurent showed that the molecules of each of these substances had the same power of deviation of polarized light. There can be no objection to such proofs.[25]

But whatever Laurent's role may have been in turning Pasteur to optical activity, Pasteur went far beyond anything Laurent had done in dealing comprehensively with Biot's general program for using optical activity to study chemical change and composition. Pasteur, like Laurent before him, was in touch with the elderly physicist while working on this thesis. The first three of its four parts harked back to Biot's own major concerns of the

---

[20] Letter to Chappuis, Paris, March 20, 1847, *Correspondance* 1: pp. 152–153.

[21] Letter to Chappuis, Paris, July, 1847, *ibid.*, pp. 154–155.

[22] L. Pasteur, "Leçons faites 1846–1847: *Notebook:* Plans de lectures de chimie faites par Pasteur probablement en 1846–1847 à l'École normale comme agrégé-préparateur de Balard," Bibliothèque Nationale, Pasteur MSS. Carton: "Lecons de chimie faites par Pasteur à l'École normale alors qu'il était préparateur de Balard, 1846–1847," 3 Folios, Folio I.

[23] Laurent was preceded here by Bouchardat's optical studies of the alkaloids. "Sur les propriétés optiques des alkalis végétaux," *Annal. de chimie* 9 (1843): pp. 213–244; "Note de M. Biot," *ibid.*, pp. 244–246; A. Laurent, "Action de quelques bases organiques sur la lumière polarisée," *Paris, Comptes rendus* 19 (1844): pp. 925–927. Laurent acknowledged Bouchardat, *ibid.*, p. 925.

Biot was something of a friend and special protector of Laurent in these years. It was he who fought for Laurent's election to the Collège de France Chair in 1850 and he wrote the admiring and moving preface to Laurent's posthumously published *Méthode de chimie*.

[24] A. Laurent, "Action des alcalis chlorés sur la lumière polarisée et sur l'économie animale," *Paris, Comptes rendus* 24 (1747): p. 220. He also tested the physiological effects of chlorinated and unchlorinated strychnine sulfate and found no difference. He took this as further proof. *Ibid.*

[25] L. Pasteur, "Thèse de physique. 1. Études des phénomènes relatifs à la polarisation rotatoire des liquides. 2. Application de la polarisation rotatoire des liquides à la solution de diverses questions de chimie," *Œuvres de Pasteur* 1: p. 20.

In Laurent lecture mentioned above in footnote 16, there is brief allusion to the use of optical activity in the study of morphine and its salts under the significant headings: "Comment les éléments sont-ils groupés?" and "Quel est le groupe particulière dans ces corps" [morphine salts].

1830's, which Pasteur characterized as being "too much neglected by chemists" in his own time, and which involved questions on the chemical states of dissolved salts.[26]

But in the fourth section, Pasteur moved on to a somewhat different subject from the chemical dynamics of the first three sections. This dealt with the hypothesis that "molecules of isomorphic substances have the same power of deviation on polarized light" and brought together, for the first time in Pasteur's research, the comparison of crystal form with that of optical activity. Moreover, Pasteur's choice of crystals was of the utmost significance here, for he chose those of tartrate salts.[27]

In fact, this fourth subject of the printed thesis is a reformulation of a somewhat different issue posed by Pasteur in an earlier written version of the thesis which Pasteur had submitted to Dumas for approval and which is signed and dated as approved on July 30, 1847. The form of the earlier version of this hypothesis, and Pasteur's commentary on it, shed light on how his ideas were rapidly evolving in the summer of 1847. The earlier formulation (there posed as the third of three general questions) was as follows:

Ought a double salt be considered as a combination of two simple salts, or is it but a simple salt whose metallic molecule contains two distinct metals?[28]

In this form, the question was more like the related questions, dealing as they did with ascertaining the true state of acids and alkalis intermixed in solutions.

In his commentary on this question, Pasteur referred to what was to be its reformulation in the published version, not, however, as something to be proved, but, on the contrary, as the basis of his argument:

Here is the principle to which I appeal for resolving this question: Molecules of isomorphic substances deviate the plane of polarization of light rays in the same amount.[29]

In support of this principle, Pasteur once again appealed to Laurent's correlation of strychnine and its chlorinated derivative with their deviation of the plane of polarized light and as a test of it, Pasteur turned to isomorphic tartrate crystals.[30] This was a perfectly natural move, considering that tartrate compounds had been the principal source of his experiments on the chemistry of dissolved optically active salts, that tartrates formed series of simple and double salts which crystallized, and finally, that these crystals had been carefully studied by De la Provostaye especially to ascertain their isomorphism.

He first tested against each other the neutral potassium and neutral ammonium salts, whose crystals "are not completely isomorphic" but which belonged to the same crystalline system and whose "modifications approach very closely; certain angles are the same or very nearly identical," and the emetic of potassium against the emetic of ammonium—"completely isomorphic."[31] Finding that their deviation was the same in each case,[32] Pasteur moved on to the resolution of the question he had posed by comparing the deviation of a double tartrate, that of potassium and ammonium with that of the neutral potassium salt, which salts were again completely isomorphic. The results were again the same, thus indicating that the structure of the double salt must be the same as that of the single salt.[33]

In the final version of the thesis, the focus shifted again to the optical activity itself, and its relation to crystalline form. In this refocusing, Pasteur once again widened the issue vastly, from the issue of optical activity and isomorphism, to the entire question of optical activity, crystal form, and molecular structure, in a remarkable prevision of the reasoning which was to lead him to his discovery of enantiomorphism the following year:

4°, Do molecules of isomorphic substances have the same power of deviation on polarized light? Tartaric acid and paratartaric acid, which act so differently on polarized light, since one of them impresses no deviation on the plane of polarization, show us how molecular arrangement can exert influence on the optical properties of a substance. By molecular arrangement, I mean here the arrangement

---

[26] L. Pasteur, "Thèse de physique," Œuvres de Pasteur 1: p. 20. In the handwritten exemplar of this thesis, Pasteur had especially inserted the phrase. Bibliothèque Nationale, Pasteur MSS. Carton: "Cristallographie et Dissymétrie moléculaire": Folio: "1847 Thèse de chimie, Thèse de physique (manuscrit)," p. 2.

[27] L. Pasteur, "Thèse de physique," Œuvres de Pasteur 1: pp. 27–30. An interesting background to Pasteur's turning to double tartrate salts is exhibited in the chemical lectures he delivered as Balard's agrégé-préparateur. Here, he invoked the isomorphism of double tartrate salts in connection with his earlier (and still concurrent) interest in the determination of atomic volumes:

"4° Les corps isomorphes ont le même volume atomique.... Je crois que la plus belle application de cette loi est relative à la détermination de la constitution des sels doubles ou acides. Ainsi les sels doubles isomorphes avec ces sels simples. Si le tartrate double de P. et d'Amm. est isomorphe avec le T.n. de Pot. ils ont le même volume atomique. Or ils ne peuvent l'avoir qu'en divisant part 2 le poids atomique du sel double." For bibliographical details, cf. above, footnote 22.

At the end of his physics thesis, Pasteur came to the same conclusion regarding the atomic weights of double salts, this time appealing to their optical activities in support of their isomorphism. "Thèse de physique" Œuvres de Pasteur 1: pp. 29–30.

[28] L. Pasteur, "Thèse de physique (manuscrit)," p. 3. Cf. above, footnote 26, for bibliographical details. Pasteur went on to say that this question had already been treated by Laurent and Dumas (the former, chemically, the latter using atomic volumes). He said that he was following Dumas's method. Ibid., p. 4.

[29] Ibid., p. 22.
[30] Ibid., p. 24.
[31] Ibid., p. 22.
[32] Ibid., p. 23.
[33] Ibid., p. 24. These were the same tartrate salts to which he had earlier appealed in his consideration of isomorphism and atomic volumes. Cf. above, footnote 27.

of atoms in the chemical molecule, since it is a matter of isomeric substances. But, aside from this arrangement of atoms in the chemical molecule, one must distinguish the arrangement of the chemical molecules themselves which, by their variable grouping, yield physical molecules of different crystalline forms. . . . Besides, as everything leads us to believe that the active power of quartz comes from the geometrical disposition of its physical molecules, and not at all from its chemical molecules taken separately or disaggregated, and that nevertheless since the phenomena relating to quartz are sensibly the same [as] in all active substances, tartaric acid excepted, I regard it as extremely probable that the mysterious, unknown disposition of the physical molecules in an entire, complete crystal of quartz is found in active substances, but this time, in each molecule in particular; that it is each molecule taken separately in an active substance, which one must compare for the arrangement of its parts to a completed crystal of quartz.[34]

Here, most of the conceptual elements necessary for the discovery of enantiomorphism were already present: the relationship between isomorphic crystal form and optical activity in the tartrates, the realization that this relationship in the tartrates was basically molecular (as opposed to quartz), and finally, the awareness that the difference in molecular arrangement in the racemates caused their difference in optical activity. The fourth (and final) element was to be the realization (and discovery) that this difference in optical activity (and molecular structure) between the racemates and tartrates must also entail differences in their crystal forms.

It is not clear whether Pasteur had this final element clearly in mind in August, 1847, when he completed his physics thesis in its final, published form. But, the studies he carried out in the intervening months between the finishing of the physics thesis and the resumption of his work on crystal form and optical activity of the tartrates and racemates in the spring of 1848, must have sharpened this issue for him, if not actually introducing it into his consciousness.

The whole cast of these intervening studies revealed the influence of Laurent more strongly than ever. The subject was dimorphism: a general survey of all dimorphic substances in order to determine their common crystallographical traits. Pasteur's guiding concept in these studies was one with which we are already familiar, since it had been enunciated and employed by Laurent. Dimorphism was not simply the manifestation of formal incompatibility, as had been thought, but rather, dimorphic crystal forms, although technically of different systems of crystalline symmetry, in fact, closely approached each other. As Pasteur put it:

. . . here is a first property common to dimorphic substances: it is that one of the two forms which they exhibit is a limiting form, a form placed in some way at the separation of the two systems, of which the one is the system proper to that form and the other, the system in which belongs the second form of the substance.[35]

What he tried to demonstrate was that this sort of relationship characterized almost all dimorphic substances. Thus, for that archetype of dimorphism, the calcite-aragonite pair, Pasteur thought he had found the means of reconciling their formal incompatibility: the forms were respectively a right rhomboidal prism whose obtuse angle was close to 120°, and a regular hexagonal prism. If one imagined the acute angles between the lateral sides of the rhomboidal prism to be truncated, the resultant form would closely approximate the "limiting" form of the hexagonal prism.[36]

In the extended version of the paper on dimorphism which Pasteur published in the *Annales de chimie*, he related his discoveries, with what was to become his accustomed historical perspective, back to the original anomaly of dimorphism to Haüy's molecular theory of crystal structure. He saw his own work as a reconciliation of this anomaly with Haüy's theory, and a vindication of Haüy's insights and the French molecular crystallographical tradition generally:

It is an error generally enough prevalent that the discovery of dimorphism dealt a profound blow to Haüy's ideas, and even that he thought that the identity of chemical composition of aragonite and calcium carbonate was not sufficiently established. It is nothing of the kind, I repeat; it is Haüy who was one of the first to insist on the dimorphism of these two substances. But here alone is where dimorphism is impossible, irreconcilable with Haüy's ideas. For many scientists, the word dimorphism means that a substance, while presenting the same crystalline geometry and the same molecular arrangement, in brief *the same substance*, could present two different crystalline forms. It is this which was inconceivable to Haüy, and, it must be said this is indeed inconceivable, at least to suppose that the molecules of one form could associate in such a manner as to give a group[ing] of another form, playing in its turn, the role of integrant molecule.

In admitting Haüy's view, one can ask why dimorphic substances do not offer difference in chemical properties as profound as isomeric substances, since isomerism and dimorphism have equally as cause a difference in the arrangement of the elementary molecules.

---

[34] L. Pasteur, "Thèse de physique," *Œuvres de Pasteur* 1: pp. 20–21. This terminology differentiating between physical and chemical molecules had been devised by Dumas in connection with the gas hypothesis. *Cf.* above, chap. 4, footnote 34, where the "molécule chimique" however, meant something like atoms rather than chemical molecule.

Ampère's later use of the terms, "particule" and "molécule" is much nearer Pasteur's here. *Cf.* A. M. Ampère, *Annal. de chimie* 58 (1835): p. 434. *Cf.* G. Delafosse, *Notice sur les travaux scientifiques de M. Delafosse*, p. 23, where he attributes the distinction in the context under discussion to Ampère and Dumas. N.B.: this is not quite the same as Delafosse's distinction in *Paris, Mém. savans étrang.* 8 (1843): pp. 649–650. *Cf.* above, chap. 4, footnote 108.

[35] L. Pasteur, "Recherches sur le dimorphisme" (*Paris, Comptes rendus* 26 [1848]: pp. 353–355) *Œuvres de Pasteur* 1: p. 35.

[36] L. Pasteur, *Œuvres de Pasteur* 1: p. 35, *Cf.* also, "Recherches sur le dimorphisme" (*Annal. de chimie* 23 [1848]: pp. 267–294), *ibid.*, pp. 43–44 for discussion of aragonite and calcite.

I share Haüy's opinion. I think that dimorphic substances are a class of isomeric ones. But if the molecular arrangements are not the same in two dimorphic varieties, there is a close relation between them. The difference is great enough to cause the incompatibility of their crystalline systems; and yet it is not at all deep. It alters the physical properties, it leaves the chemical properties almost the same.[37]

At the end of the short version of this paper which Pasteur presented to the Académie des Sciences on March 20, 1848, he made two remarks of the greatest significance in placing this work on dimorphism in the context of his development towards the discovery of enantiomorphism. The first looked back to Laurent: Pasteur saw his work on dimorphism as offering support to Laurent's concepts of hemimorphism and to his widened definition of isomorphism which he made in 1845.[38] The second comment looked forward to his breakthrough to enantiomorphism: Pasteur, working with two students of the École Normale, had extended his concept of dimorphism to the tartrate salts. Pasteur claimed to have found that eight tartrates, the neutral tartrates of potassium, sodium, and ammonium, the double tartrates of potassium and ammonium, of potassium and sodium, and of sodium and ammonium, and the bitartrates of potassium and of ammonium, approached so closely in their crystalline forms as to be considered isomorphic, crystallizable together in any proportion, even though some of them crystallized in a system technically incompatible with that of the others.[39]

With this, the pieces of the solution to the puzzle of the tartrates and the racemates were almost all in place. The conceptual advance Pasteur had made over his physics dissertation was one of generalization; in the dissertation, he had shown that those tartrates which were isomorphic were also identical in their optical activity. Carrying on his studies of dimorphism under the influence of Laurent's ideas, he was led to show that all the tartrate salts were isomorphic, in the wider sense of the term as devised by Laurent in 1845. Pasteur had succeeded where De la Provostaye had failed, in fitting the tartrates to the Procrusteean bed of isomorphism—by lengthening the bed.

Consideration of the isomorphism of these eight tartrates became the context, or better the point of departure, for Pasteur's discovery of enantiomorphism. We are in an extraordinarily good position to follow the steps which led to this discovery because the pages of Pasteur's laboratory notebook, in which he recorded his experiments and his thoughts at just the time of this discovery, have survived. These pages were examined by J. D. Bernal in his study of Pasteur's discovery made to commemorate its centenary a quarter of a century ago. But Bernal's examination was, by his own admission, cursory; moreover, he appears to have made little study of Pasteur's researches prior to that resulting in the great discovery and hence was unaware of the significance of the work on dimorphism and isomorphism.[40]

The very first page of the crucial sequence, a photocopy of which Bernal published in his study, bears out my point about the contextual significance of the isomorphism of the tartrates. This page has a title: "Tartrates (questions à resoudre)" and consists of a listing of the eight tartrates which Pasteur had asserted were isomorphic, along with their chemical formulas. Beneath this list, these tartrates are subdivided into smaller groups of two or three, on the basis of whether they have the same "type" formula.[41]

The point at issue here seems to have been a disparity which bothered Pasteur, between his alleged isomorphism of all these tartrates and their chemical formulas. They differed among themselves in the amounts of water of crystallization their molecular formulas contained. Some seemed to contain one molecule of water, some four, and some eight. But from the

---

[37] *Ibid.*, pp. 38–39. He added that he had initially believed that dimorphism had serious consequences for Haüy's theory, but that Chevreul had set him right, asserting to him that Haüy had never said that substances of the same composition had to have the same form. *Ibid.*, p. 39, footnote 1.

[38] L. Pasteur, "Recherches sur le dimorphisme," *ibid.*, p. 37. Cf. also "Note sur un travail de M. Laurent intitulé: sur l'isomorphisme et sur les types cristallins" (*Annal. de chimie* 23 [1848]: pp. 294–295), *ibid.*, pp. 59–60.

[39] L. Pasteur, "Recherches sur le dimorphisme," *ibid.*, p. 37. He saw this as "un preuve de plus en faveur des idées qui sont la base de ces recherches" [i.e., the Laurentian ideas behind Pasteur's work on dimorphism]. *Ibid.*

[40] J. D. Bernal, *op cit.*, pp. 195 ff. He wrote concerning them "I saw them only a day before this lecture; they deserve a detailed and thorough study." *Ibid.*, p. 195, footnote 2.

[41] I must thank Professor Wyart of the Laboratoire de Minerologie-Cristallographie, Université de Paris, for his great kindness in making microfilms of these laboratory notebooks available to me at a time when the notebooks themselves were inaccessible. All subsequent quotations are taken from these notebooks. I have omitted passages which are crossed out unless they add to the meaning significantly.

The first page (also reproduced in *ibid.*, p. 196) has, below the eight tartrates and their formulas, the following division:
"Parmi ces sels, voici ceux dernière formule type:
Tartrate neutre d'Ammoniaque.
{ bitartrate d'Ammoniaque
{ bitartrate de Potasse } isomorphes (de même syst.)
Le Tartrate neutre d'Ammoniaque est isomorphe avec le bitartrate
d'Ammoniaque bien que le système soit différent.
x Faire cristalliser le T. neutre d'Amm. et le bitartrate d'Amm.
x Répéter l'expérience qui a donné du bitartrate d'Ammon, en chauffant un mélange de T. neutre d'Amm. et du Tart. double d'Ammon et de soude.
Autres de même formule:
{ Tartrate neutre de Potasse
{ Tart. double de Pot. et d'Amm. } isomorphes (même système)
Sel Seignette et Tart. double d'Ammon. et soude."
The marginal "x" was Pasteur's method of indicating something to be researched.

time of Mitscherlich's first work on isomorphism, water of crystallization had been considered an integral component of isomorphic formulas. As Pasteur summarized the problem in his notes:

> Since it is quite probable that these formulas are all exact, the isomorphism of these salts is in need of confirmation. For, if it really exists, and if it is not accidental that there is a relationship between the forms of these salts, it would be necessary that these salts crystallize together.[42]

Or, as he put it after proposing the investigation of four such mixtures of disparate tartrates:

> —all these salts have different type formulas—do they nevertheless crystallize together? This is the question to resolve.[43]

One particular case led Pasteur farther afield from the tartrates, to the racemates and to Mitscherlich's note of 1844: that of the double sodium-ammonium tartrate compared with the double sodium-potassium tartrate (Seignette salt). Pasteur had bracketed these two salts together as having perfectly isomorphic formulas (each with eight molecules of water of crystallization). But, as he admitted, he had assumed that the formula of the sodium-ammonium salt was of the same type as Seignette salt because of their apparently complete crystallographical isomorphism. This, however, raised a problem of decisive import:

> I shall remark on this subject that M. Mitscherlich has said that the double paratartrate of sodium and ammonium and the double tartrate of these bases were completely isomorphic and of the same formulas; and that the one, however, did not deviate the plane of polarization. But M. Gerhardt gives 2 HO for the quantity of water of the paratartrate. Would that also be the quantity of water of the double tartrate [?] But then how to explain its isomorphism with Seignette salt? There is something here to be reviewed.[44]

Then came the fateful plan of research:

> Separate the double paratartrate of sodium and ammonium. Its form and analysis? idem. analyse the d(ouble) T(artrate) of these bases. Ask M. Biot if he still has a sample of the double paratartrate which M. Mitscherlich sent to him and see if it is isomorphic with the corresponding double tartrate or with Seignette salt.[45]

We note here that there is nothing yet about the isomerism of the tartrate and racemates, although Pasteur, as evidenced in his physics thesis, was quite aware of this.

But then occurred a significant entry, the first mention by Pasteur of hemihedry in tartrates. It was not the sodium-ammonium tartrate or the Seignette salt, however, but crystals of two other tartrate salts to which he must have then turned: the neutral potassium and the neutral ammonium tartrate. But the hemihedry of these salts was mentioned almost incidentally. Pasteur was still worrying about isomorphism; he was concerned over the disparity between the crystallographical isomorphism of these two salts, now even to their hemihedry, and the difference in their amounts of water of crystallization.[46]

However, almost at once, Pasteur made the connection with matters other than mere chemical formulas. He recalled an observation made by W. Hankel in 1843 on the hemihedry and pyroelectricity of potassium tartrate crystals.[47] It was, no doubt, a tribute to Delafosse's impact on Pasteur, that he should have noticed Hankel's observation in the first place and remembered it now. Pasteur suggested testing the ammonium tartrate for pyroelectricity.

---

[42] "Comme il est très probable que ces formules sont toutes exactes l'isomorphisme de ces sels a besoin de confirmation. Car s'il existe réellement et si ce n'est pas un fait accidental que la relation des formes de ces sels, il faudra que ces sels puisse cristalliser ensemble."

[43] Neutral potassium tartrate with double sodium-potassium tartrate; neutral sodium tartrate with double sodium-potassium tartrate; neutral ammonium tartrate with double sodium-potassium tartrate; double sodium-ammonium tartrate with double potassium-ammonium tartrate, were the proposed mixtures:

"—Tous ces sels ont les formules types différentes—ont-ils cependant cristallisé [sic] ensemble? C'est la question à resoudre."

[44] "Parmi les formules citées plus haut celle du T.d.d'Am. et S. n'est donneé nulle part. Je l'ai regardée comme la même que celle du sel Seignette à cause de l'isomorphisme complet de ces 2 sels.

Je remarquerai à ce sujet que M. Mitscherlich a dit que le paratartrate double de Soude et d'Ammoniaque et le T.d. de ces bases étaient complétement isomorphes et de même formule et que l'un cependant ne deviait pas le plan de polarisation. Or M. Gerhardt donne 2 HO pour la qté d'eau du Paratartrate. Serait-ce là la qte d'eau du T. double [?] Mais alors comment expliquer son isomorphisme avec le sel Seignette[?] Il y a là quelque chose a revoir."

J. D. Bernal, op. cit., p. 197, somewhat garbled this passage in translation.

[45] "x Séparer le Paratartrate double de S et Am. Sa forme son analyse? idem. analyser le T. d. de ces bases. Demander à M. Biot s'il aurait encore un échantillon du Paratartrate double que lui avait remis M. Mitscherlich et voir s'il est isomorphe du T.d. correspondant ou du sel Seignette."

[46] Notes sur les Tart. n. de Pot. et d'Am.

1. Ils sont tous deux hemyédrés.
2. Ils sont isomorphes ou du moins les deux formes quoique différentes sont compatibles.
3. Ils possedent tous deux un clivage facile parallélment à la même direction.

Cependant M. Dumas ne leur attribue pas la même formule. x Les analyses de ces sels sont donc à reprendre. Reprendre d'abord la qté du Tart.n. par une analyse organique—

[47] Note sur le Tartrate neutre de Pot.

M. Hankel (3 lannee 1843 Ctee Rdus de Berzelius, p. 135) observe que le T.n. de P. est hémyédre et possède l'électricité polaire quant on élève la tempre. En élevant la tempre une des extremites est positive et elle est négative quand la sel se refroidit.

Le t.n. d'Amm est aussi hémyédre aux extrémités correspondantes à celles du T.n. de Pot. Une certtine cristallisation a donne de forts beaux cristaux de T.n. d'Am. où cette hémyédrie est [ou] ne [point] plus evidente. Essayer s'il jouit de [l'électricité] polaire.

On the following page, we witness what must have been the dawning realization by Pasteur that all the tartrates are not only crystallographically, at least, isomorphic, but also hemihedral. For the heading of this page is "Notes sur l'hémyhédrie des Tartrates en général' and four cases of hemihedry are listed.[48] Three pages further on, dated for the first time "29 avril, 1848," Pasteur came at last to the sodium-ammonium tartrate and found it hemihedral. Three pages beyond this is found the examination of the corresponding racemate, the discovery of its left- and right-hemihedral varieties, one of which is identical to the tartrate, and finally, the testing of them in solutions and his demonstration of optical activity. With this, the great discovery was achieved.[49]

The guiding thread in all of this was the isomorphism of the tartrates, the conclusion of his previous investigation. His attempt to refine his allegation about their isomorphism by considering their chemical formulas was the impetus to his taking up the question of the salts discussed by Mitscherlich in 1844; and there is little doubt that, once he discovered hemihedry in a few tartrates, the consideration of isomorphism led him to look for it in them all. Finally, in his initial interpretation of his discovery, it was the same isomorphism which played the key role. Indeed, in this interpretation, Pasteur made the discovery of enantiomorphism appear as even more of a logical follow-through to his ideas on the dimorphism paper than it in fact had been.

We are again fortunate in possessing what must surely be the first draft of the note Pasteur wrote to the Académie des Sciences announcing his discovery. The argument of this note quite possibly represents the stage of Pasteur's thoughts just after he had found the sodium-ammonium tartrate to be hemihedral and was about to turn to the racemates;[50] it certainly represents Pasteur's first reflections on the significance of his discovery.

He began his argument with a reassertion of his allegation that all the tartrates are isomorphic (in the wider sense of Laurent); they differ from one another only in the modifications of their extremities, their angles and edges. This, he put in terms of a "nuclear" molecule structure model strikingly reminiscent of Laurent:

I think from this that it is impossible to doubt that a certain molecular group remains constant in all these salts; that the water of crystallization, the bases, relegated to the extremities of this group, modify it at the extremities only, hardly touching the central molecular arrangement and there only in the difference of angles observed between the facets.... Moreover these facts show us, in addition, the close relationship which exists between crystalline form and molecular constitution and the day when one can discover from crystallographical studies the arrangement of atoms.[51]

By this model, then, Pasteur was able to incorporate, generally, the disparities in the amount of water of crystallization of the tartrates. But if the tartrates were isomorphic, so must be the racemates and, moreover, not only with themselves but also with all the tartrates:

In such a way there exists a molecular group common to all the paratartrates and this group is the same as (that) in the tartrates.[52]

If this be the case, then the peculiarity of Mitscherlich's original sodium-ammonium tartrate and racemate, with their identity of composition and crystalline form but isomerism and disparity in optical activity, was generalizable to all tartrates and racemates. There simply had to exist some fundamental difference between the molecular, and hence crystal, forms of tartrates and racemates if the anomaly was to be resolved. I quote in extenso the key passage from Pasteur's un-

---

[48] *Ibid.*
— T.n. de Pot.
— T., d'Am.
— Emétique d'Am. et de Pot.
— Emétique d'Am à plusieurs eqts d'eau (?)

[49] *Cf.* J. D. Bernal, *op. cit.*, pp. 200-206 for a careful and moving account of these discoveries, with copious citations from the notes.

[50] Pasteur did write much later that, after discovering hemihedry of the tartrates, he had expected *not* to find it in the paratartrates, and thus "la note de Mitscherlich n'aura plus de mystère, la dissymétrie de la forme du tartrate correspondra a sa dissymétrie optique; l'absence de dissymétrie de la forme dans le paratartrate correspondra a l'inactivité de ce sel sur le plan de la lumière polarisée, à son indifférence optique." But, when examining the sodium-ammonium paratartrate, "j'eus un instant un serrement de cœur; tous ces cristaux portaient les facettes de la dissymétrie." "La dissymétrie moleculaire" (Conférence faite à la Société Chimique de Paris le 22 Decembre 1883), *Œuvres de Pasteur* 1: p. 371.

[51] "Je pense dès lors qu'il est impossible de douter qu'un certain groupe moléculaire reste constant dans ces sels, que l'eau de cristallisation, que les bases reléguées aux extrémités de ce groupe le modifient à ces seulement ne touchant qu'à peine et dans la mesure de la différence des angles observés entre les facettes, à l'arrangement moléculaire central.... Ces faits cependant nous montrent en outre l'étroite relation qui existe entre la forme cristalline et la constitution moléculaire et le jour que l'on peut jeter par les études cristallographiques sur l'arrangement des atomes."

Just before this passage, Pasteur had written: "Les formes pourront appartenir à des systèmes différens et a côté du Prisme Rhomboïdal on pourra trouver le prisme rectangulaire droit ou oblique ou même le prisme tout à fait oblique du dernier système cristallin mais méamoins les angles des pans ou ceux des facettes de modification différont très peu les uns des autres. Quand deux formes primitives ne seront pas du même système, l'une sera pour l'autre une forme limite. Je fais ici abstraction des extrémités des prismes, et en effect c'est par les extrémités seules des prismes que différent les formes cristallines de tous les Tartrates."

[52] "De telle sort qu'il existe un groupe moléculaire commun à tous les Paratartrates et que ce groupe est le même que dans les Tartrates."

published draft, with its elaboration of the Laurentian molecular model:

> If the tartrates and paratartrates offer between them such relations, one might ask how it can be that there is isomerism between these two series of salts. We see, in effect, that the crystalline forms of all the tartrates only differ between themselves, and from the paratartrates, by the extremities of [their] prisms [and] that it is the same w[ith regard to] the paratartrates compared with themselves and with the tartrates. If one compares all the tartrates, as I have said, the extremities alone of their forms are different. But if one considers a particular tartrate, one will soon see that the two extremities of the prism are disymmetrical in this tartrate, without there remaining any doubt in this regard. The law of the famous Haüy, which posits that identical parts should be modified in the same manner, is violated. All the tartrates are hemihedral. Thus, the molecular group common to all these salts, and which the introduction of water of crystallization and of oxides comes to modify at the extremities, does not receive the same element at each extremity, or, at least, they are distributed in a disymmetrical manner. On the contrary, the extremities of the prism of the paratartrates are symmetrical. Behold all the difference. It suffices to explain the isomerism of these two types of salts, and, at the same time, we see that this isomerism is not very profound. There is a difference in the molecular arrangement of tartrates and paratartrates, but this difference seems only to stem from a regular distribution of the oxide or water of crystallization molecules in the one case and a disymmetrical distribution in the other.[53]

Pasteur then proceeded to the heart of the case—left- and right-hemihedral paratartrates:

> ... if there existed a paratartrate which was hemihedral and which possessed the guise of the tartrate's hemihedry, one could only conceive that this paratartrate was a veritable tartrate. And—most extraordinary—this paratartrate exists. There exists a salt prepared with pure paratartaric acid which is hemihedral and which possesses, moreover, all the physical and chemical properties of the corresponding tartrate. And, even more extraordinary, this same paratartrate, rather than being hemihedral in the manner of all the tartrates—hemihedral to the left—can be hemihedral to the right and still is a veritable tartrate. Hemihedral to the left, it deviates the plane of polarization as all the tartrates do; hemihedral to the right, it deviates the plane of polarization to the left.[54]

The general argument and some, if not most, of the references to a nuclear molecule model, were incorporated in the actual note to the Académie announcing the discovery.[55] But Pasteur suppressed it all in the first extended paper he published on enantiomorphism and he never again alluded in print either to the argument or to the model.[56] Moreover, in his own later

---

[53] "Si les Tartrates et les Paratartrates offrent entre eux de pareilles relations on se demande comment il se peut qu'il y ait isomérie entre ces deux séries de sels. Nous voyons en effet que les formes [cristallines] de tous les Tartrates ne différent entre eux et des Paratartrates que par les extrémités des prismes, qu'il eu est de même des Paratartrates comparés entre eux et aux Tartrates. Voici la différence essentielle. Si l'on compare tous les Tartrates, ai-je dit, les extrémités seules des formes seront différentes. Mais si l'on considére un Tartrate en particulier on verra bientôt san qu'il reste ombre de doute à cet egard que dans ce Tartrate les deux extrémités du Prisme sont dissymétriques. La loi du célèbre Haüy qui veut que les parties identiques soient modifiés de la même manière est violeé. Tous les Tartrates sont hemyédriques. Ainsi ce groupe moléculaire commun à tous ces sels et que l'introduction de l'eau de cristallisation, et des oxydes vient modifier aux extrémités ne reçoit pas à ces deux extrémités les mêmes élémens, ou du moins ils y sont distribués d'une manière dissymétrique. Au contraire dans les Paratartrates les extrémités du prisme sont symétriques. Voila toute la différence. Elle suffit à ous rendre compte de l'isomérie de ces deux genres de sels, et en même temps nous voyons que cette isomérie n'est point cependant très profonde. Il y a différence dans l'arrangement moléculaire des Tartrates et des Paratartrates, mais cette différence ne parait provenir que d'une distribution dans un cas regulière dans l'autre cas dissymétrique des molécules d'oxyde ou d'eau de cristallisation."
The passage from "Si les Tartrates . . ." to "Ainsi ce groupe . . ." is lightly crossed out but still legible. The same is true of the passage beginning with "Au contraire . . ." and going through ". . . nous voyons cette isomérie . . ." (ending with the word, "isomérie").

[54] ". . . s'il existait un Paratartrate qui fut hémyédre et qui possedât le guise d'hémyédrie des Tartrates, on ne concevrait pas que ce Paratartrate ne fut pas un véritable tartrate. Chose extraordinaire ce Paratartrate existe. Il existe un sel préparé avec l'acide Paratartrique pur qui est hémyédre et qui possède alors toutes les propriétés Physiques et chimiques du Tartrate correspondant. Chose bien plus étrange! Ce même paratartrate au lieu d'être hémyédre comme tous les Tartrates, hémyédre gauche, peut être hémyédre à droite et alors c'est un véritable Tartrate [although he has written "Paratartrate," he has gone over the "T" in "Tartrate" clearly to show that this is what he meant, and indeed, the sense of the passage requires "tartrate" rather than "paratartrate."] Hémyédre à gauche il devie le plan de Polarisation à droite comme tous les Tartrates, hémyédre à droite il devie le plan de Polarisation à gauche."
From "Chose extraordinaire . . ." the passage is lightly crossed out but still legible.

[55] The passages quoted in footnotes 51 and 52, for example, were retained in the note to the Académie. L. Pasteur, "Mémoire sur la relation qui peut exister entre la forme cristalline et la composition chimique, et sur la cause de la polarisation rotatoire" (*Paris, Comptes rendus* [seance du 15 mai, 1848] **26**: pp. 535-538), *Œuvres de Pasteur* **1**: pp. 61-62.

[56] There is a draft of the first paper on enantiomorphism published in the *Annales de chimie* ("Recherches sur les relations qui peuvent exister entre la forme cristalline, la composition chimique et le sens de la polarisation rotatoire" [*Annal. de chimie* **24** (1848): pp. 442-459], *Œuvres de Pasteur* **1**: pp. 65-80)—one of a number of such exemplars of this paper—in which Pasteur had included an extended passage in the introduction very much like the one quoted in footnotes 53 and 54. The ending of this passage is particularly interesting in its reference to the context of Laurent's work: "De pareilles recherches pourraient éclairer sans aucune doute bien des questions de classification en Minéralogie. La comparaison des formes des silicates ne peut manquer d'offrir des relations analogues à celles des Tartrates et des Paratartrates et on parviendrait peut être de cette mécanisme à les classer en groupes naturels. On ne peut supposer que le fait que je signale ici pour les Tartrates soit tout à fait isolé. Si l'on jette les yeux sur les formes cristallines du nombreux produits: derivés de la Naphtaline études par M. Laurent on ne tarde pas à reconnaître que dans ces combinaisons, même celles où le type est tout à fait différentes [sic] en apparence certains angles des pans dans les cristaux sera tout constamment. Ce fait n'a pas échappé à la sagacité de M. Laurent et il n'a pas hesité à affirmer qu'il devait y avoir dans toutes ces substances un groupe moléculaire invariable." Bibliothèque

accounts of how he had come to make the discovery, which have been uncritically accepted, Pasteur telescoped together his first encountering of Mitscherlich's note in his student days at the École Normale with the discovery of enantiomorphism itself.[57] The intervening researches, his chemistry thesis, his physics thesis, and his work on dimorphism, were all forgotten. The story that emerged, as so often happens in retrospection, is a very personal one, almost without context. Pasteur had been struck by the anomaly of the tartrates and racemates in Mitscherlich's note, this had puzzled him, it continued to bother him, he investigated it and he removed the anomaly with the discovery of enantiomorphism.

Yet, as I have tried to show, close attention to his research interest prior to the discovery and to the steps of the discovery itself, reveals a more complicated story, and, more significantly, one with a context in French crystallographical chemistry. It is, I hope, possible now to appreciate that, if ever a mind was "prepared" for a great discovery, it was Pasteur's, by French molecular crystallographical chemistry as exemplified by the work and ideas of his mentors, Delafosse and Laurent.

CONCLUSION

If my reconstruction of the steps in Pasteur's discovery of enantiomorphism is correct, this achievement was a particularly logical and even inevitable fruition of the activities of the French scientists with whom I have been dealing in this work. The theme which ran through their work also informed Pasteur's discovery: the importance of molecular form as a material attribute and the relevance of crystallography for elucidating this form. As I have tried to show, this was a theme which went back to Haüy's theory of crystal structure. With his conception of the *molécule intégrante* as a physical as well as mathematical entity and with his concern with the chemical implications of his theory, at least for mineral classification, Haüy himself had set the stage for the subsequent attempts to elucidate molecular form through the union of crystallography with chemical atomism.

I have referred to these attempts as the "activities" of a number of French scientists, rather than more forcefully, as a "tradition." I do this with circumspection because I am well aware that what I have described, however great its importance for nineteenth-century physical science, does not fall easily into the category of "normal" scientific enterprise and, indeed, has been completely submerged from historical view and identification until very recently. The men involved were very few in number; their institutional, pedagogical, and professional interconnections were tenuous, when their personal relations were not actually antagonistic. Haüy's molecular theory of crystal structure served in a general sense as the background to their speculations, and, more particularly, Ampère's conception of a geometrical chemistry, as outlined in the 1814 paper, served as something like a unifying paradigm for their approach. But it is difficult to find much in the way of a cumulative and systematic development of Ampère's program in their work.

Yet, as I have tried to show, the common features of their speculations came to a focus in the achievement of Pasteur through the influences of Delafosse and Laurent. It was the combination of Delafosse's principle that crystalline hemihedry reflected molecular dissymmetry with Laurent's widened conception of isomorphism, which the latter had made use of in the mid- and late-1840's in order to extend his nuclear theory to the mineral realm, that made it possible for Pasteur both to apprehend the enantiomorphism of the tartrates and immediately to understand its significance. Without a knowledge of this background, Pasteur's discovery remains astonishing and inexplicable. With it, the discovery becomes part of the context of French interest in crystallographical-chemical speculation.

There remains one unanswered question in my reconstruction: why did Pasteur fail to give due credit to Laurent if his was, as I have argued, the crucial influence on Pasteur? Pasteur was, after all, a scientist with at least the pretense to an unusually sensitive historical perspective and one indeed who wrote extensively on the relationship of his discovery to the general development of French crystallography. He was properly appreciative of Haüy's seminal role in the development of this science and he gave Delafosse credit for his influence. This question is not an easy one to answer and I have been able to discover only one concrete piece of evidence bearing on it, but this is, at least, a highly suggestive one. It is a letter which Pasteur wrote to Dumas in February, 1852. In the course of a fulsome expression of gratitude to both Dumas and Biot for their guidance, Pasteur recounted briefly his own development and mentioned some of his mentors. In particular, he mentioned Laurent:

I had worked under the guidance of the good M. Laurent whom death is, perhaps, soon to remove from science. I was at an age when the spirit is fashioned on the model which is presented to it. I was enveloped by hypotheses without basis, by a redaction which completely lacked pre-

---

Nationale, Pasteur MSS. Carton: "Cristallographie et Dissymétrie moleculaire": "Premier projet de travail sur les relations qui peuvent exister...."

The entire passage was omitted from publication.

[57] E.g. in his "Recherches sur la dissymétrie moléculaire des produits organiques naturels" (Leçons professées à la Société Chimique de Paris le 20 janvier et le 3 fevrier 1860), "Premier Leçons, *Œuvres de Pasteur* 1: pp. 323–324. He did write that he was guided especially by the views of Delafosse ". . . pour qui l'hemiédrie a toujours été une loi de structure et non un accident de la cristallisation" to correlate the hemihedry of the tartrates with their optical activity. *Ibid.*, p. 372.

cision, and I spoiled the exposition of new and interesting facts. I was quickly enlightened by your counsel.[1]

In this passage, the disavowal of Laurent is oblique, but nevertheless seems clear. As to what were the "hypotheses without basis," Pasteur did not elaborate. But from my reconstruction, this presumably referred to the extended conception of isomorphism which Pasteur had adopted from Laurent, and even more specifically the sketches of molecular nuclear models like those of Delafosse and Laurent, which were suppressed in the published version of the paper on enantiomorphism. Neither the isomorphism nor the nuclear models appear again in Pasteur's discussions of enantiomorphism; if our interpretation of the passage from the letter is correct, then Pasteur had come to reject them.

The context of this strange and oblique disavowal of Laurent, in a letter to Dumas, might raise suspicions about Pasteur's integrity. But from whatever motive it stemmed, Pasteur's repudiation of his "hypothesis without basis" was expressive of a scientific truth of his times: that the attempt at synthesizing crystallography and chemistry to explicate atomic arrangement and molecular form had reached the limit of its potentiality, for the time being, in Pasteur's own work. To proceed further, to determine the specific arrangements of atoms in molecules by crystallographical-chemical means, as Pasteur himself seems originally to have believed possible, was not possible at this time, given the state both of chemistry and crystallography.[2] Indeed, neither Laurent, nor Delafosse nor, for that matter, Pasteur succeeded in anticipating the breakthrough to stereochemistry of Van't Hoff and LeBel in 1874. The direct inferral of specific atomic arrangements from crystal structure was not to be achieved until the twentieth century; the rise of organic stereochemistry was the result of a more indirect, purely chemical route and was made possible only by the advances in chemistry of the two decades subsequent to Pasteur's discovery: the development of structural chemistry, chemical valence theory, and the correct computation of atomic weights and formulas through the use of the gas hypothesis.

Many of these developments had been previsioned or initiated by our scientists in the 1830's and 1840's and, while much more detailed study of the period subsequent to 1848 is still required, it has already become apparent how very significant was Laurent's general structuralism for the rise of structural chemistry at the hands of Kekulé, Butlerov, and others. But the more specific attempts to explicate atomic arrangement through the synthesis of crystallography and chemistry failed to catch on. Pasteur's discovery, then, like all great achievements in science, was Janus-faced. At the time, it was a climax and denouement to what had been a fascinating but premature attempt to found a geometrical chemistry. In the long run, it was to be the basis for the establishment of just such a chemistry. In the interim years, the discovery of enantiomorphism remained in a kind of honorable limbo. Honorable, because Pasteur received instant recognition for having made a significant discovery, but a limbo nevertheless because, although its general significance was clear, the specific implications were not understood until developments in organic chemistry had produced the means of unraveling molecular dissymmetry in terms of specific atomic arrangements.

Pasteur himself continued to follow up the ramifications of his discovery for almost another decade, and it was this work which largely provided the foundations for stereochemistry. But for Pasteur, the main fascination of molecular dissymmetry came to lie not in the further elucidation of atomic arrangement, but rather in what it seemed to indicate about the disjunction between life and non-life. It was a kind of structural chemical vitalism, in which molecular dissymmetry could only be produced by the activities of living organisms. This idea came to fascinate Pasteur and eventually turned him towards medical biology and the germ theory of disease. Thus, just as Pasteur's discovery of enantiomorphism itself climaxed a chapter in speculative structural chemistry opened by Haüy's molecular theory of crystal structure, so there is a neat symmetry here between Haüy's original turning to mineralogy from botany in order to discover the kind of morphological interrelations among rocks and stones that Linnaeus had found in plants, and Pasteur's seeing the distinguishing characteristic of life in molecular dissymmetry, itself inferred from crystallographical enantiomorphism.

---

[1] Letter to J. B. Dumas, Strasbourg, February 5, 1852, *Correspondance* 1: p. 236.

[2] While Pasteur never proposed any specific models of atomic arrangement, nevertheless the notion of a "molecular grouping" common to different but closely related substances and manifested by their correspondence in optical, crystallographical, and chemical properties, played an important role in the development of his ideas on the relationship between malic and tartaric acid. Cf. D. Huber, "Louis Pasteur and Molecular Dissymmetry: 1844–1857 (M.A. thesis, John Hopkins University, 1969), pp. 44 ff. Ms. Huber provides a detailed account of Pasteur's investigations subsequent to his great discovery of 1848. The closest Pasteur came to suggesting a specific molecular model was in a now-famous passage from a lecture which Pasteur delivered to the Société chimique de Paris early in 1860. He speculated on whether the atoms in the *dextro*-tartaric acid molecules might be arranged along a right-turning helix or placed at the summits of an irregular tetrahedron, but commented that the answer could not yet be given as to which model was the correct one. "Recherches sur la dissymétrie moléculaire," *Œuvres complètes* 1: p. 327.

INDEX

Académie des Sciences, 9, 37, 40, 41, n. 44, 48, 54, 57, 58, 62, 67, 72, 75, 77, 78
Alloys, 60
Ampère, André Marie, 33, 35–37, 38, 39, 40, 42, 46, 50, 51, 52, 53, 54, 55, 79
Aniline, 72
Arago, Dominique François Jean, 56, 58–59, 61
Aragonite, 30–31, 49, n.93, 58–59, 74
Arcueil, Société d', 59
Arsenious acids (and compounds), 71
Atomic divisibility, 46, 50
Atomic rows, 40–41 (fig. 2)
Atomic theory, 5–7, 21, 26–28, 31–33, 35, 41, 42, 53, 57, 79
Atomic volumes, 72, 73, n.27
Atomic weight, relative, 5, 6, 26–27, 42, 50, 80
Atoms, 27–28, 31, 33, 34–37, 39, 41–42, 46, 50, 55
d'Aubuisson de Voisins, Jean François, 23–24
Avogadro, Amedeo, 35, 38

"Balance of nature," 22
Balard, Antoine Jérome, 68–69, 71
Barlow, William, 35
Bartholin, Erasmus, 55
Baudremont, Alexandre Édouard, 37, 40, 41, 42–45, 46, 47, 48, 50, 51, n.105, 54, 55
Becquerel, Antoine César, 39
Bergman, Torbern Olof, 7, 11–12, 14
Bernhardi, Johann Jacob, 13, n.35, 29, n.42, 57–58
Berthollet, Claude Louis, 21, 21–26, 28, 35, 59, 60, 65
Beryl, 28–29
Berzelius, Jöns Jacob, 27–28, 32, 41–42, 66–67, 72
Beudant, François Sulpice, 29–30, 60
Biot, Jean Baptiste, 6–7, 56, 58–62, 63, 64–68, 69, 70, 72, 76
Boracite, 34, n.9, 52–53 (figs. 3, 4)
Bouchardat, Apollinaire, 72, n.23
Bravais, Auguste, 6, 33, 51, 52, 53, 54, 55
Brewster, David, 56, 58, 65, n.63
Brongniart, Alexandre, 26
Brooke, Henry James, 13, n.35
Buffon, George Louis Leclerc, Comte de, 15–16

Calcspar (calcareous spar, calcite, Iceland spar), 7, 11, 16, 29, 30–31, 49–50, 56, 58–59, 61–62, 74
Camphor, 43, 62
Cane sugar, 41, fig. 2, 62
Cannizzaro, Stanislao, 38
Carangeot, Arnould, 10, n.19
Chappuis, Charles, 71–72
Chemical affinity, 8, 17, 21–22, 24–25, 38, n.31, 58–59
Chemical analysis, 18–19 (and mineral classification)
Chemical combination, 5, 22, 24, 35–37, 40, 41, 43, 59–60; law of definite proportions, 20–22, 23, 26–27, 29, 32, 34–35, 61; law of multiple proportions, 21, 27–28, 32, 33–35; in variable proportions, 20–21, 23–26, 60, 65–66; "planetary theory" of, 38, n.31, 49, n.95, 65
Chemical composition, 10, 17–19, 32 (of fixed mineral species); 28–32 (isomorphism and polymorphism); 56, 58, 60, 65 (determined by optical means); 66–67 (isomerism)
Chemistry, 5–7, 10, 20; 25–26, 28, 32 (in mineral taxonomy); 35–51 (in the programs of Ampère, Gaudin, Baudrimont and Laurent); 54–55 (in Delafosse); 56, 58–60, 65–67 (vs. physical optics and crystallography in elucidation of molecular composition); 69, 71–73 (Pasteur); 80; organic, 5, 6, 38, n.31, 41–42, 44–45, 48–50, 54
Chenevix, Richard, 26
Chevreul, Michel Eugène, 75, n.37
Cinchonine, 72
Cleavage, 11–12
Collège Cardinal Lémoine, 7
Collège de France, 37, 38, 39
Condillac, Étienne Bonnot, Abbé de, 18
Condorcet, Marie Jean Antoine Nicolas Caritat, Marquis de, 21, n.2
Constant interfacial angles, law of, 10, 17, n.21
Cronstedt, Axel Fredrik, 10, n.17, 14
Crystal form, 7–12, 13, fig. 1, 14–16, 18–19, 23, 25, 26, 28, 29–30, 32, 40, 43, 47–48, 52–53, 54–55, 57, 59, 61, 63, 66–68, 69, 70, 71, 73, 74, 76, 77
Crystal lattice structure, 6, 36, 51, 52, 55
Crystalline axis, 58
Crystallization, 8, 10, 25, 29, 33
"Crystallographical-chemical approach," 33, 38–53, 68, 71, 79–80
Crystallographical systems, 40, 49, 52, 54, 74–75
Crystallography, 5–7, 10–11; 19–20, 25–26, 32 (vs. chemistry in mineral species determination); 40, 45, 50–51, 53 (Gaudin, Laurent, Delafosse); 56, 58–59 (vs. chemistry & physical optics in elucidation of molecular form and composition); 59, 79–80 (as background to Pasteur)
Crystals: nucleus (noyau), 12; primitive form, 9–12, 28, 30, 35–36, 43, 52–53, 57–58, 70; secondary form, 11–12, 53–54
Crystal structure, molecular theory of, 5–14, 17–19, 30, 33–35, 40, 42, 48, 51–55, 56–58, 63, 70, 74, 79–80

Dalton, John, 5, 6, 7, 20, 26–27, 28, 31, 33, 60
Dana, James Dwight, 35
Daniell, John Frederic, 35
Daubenton, Louis Jean Marie, 16–17, 18, 20
Davy, Humphry, 35, 37
Decrements, molecular, 12, 13, fig. 1, 53, 57, 58, 63

Demeste, Jean, 11, n.22
Depolarization, 61
Dextrin, 66
Dolomieu, Deodat de Gratet de, 14, 17, 19–20, 21, 23, 24, 25
Dolomite, 50
Delafosse, Gabriel, 7, 33, 41, n.44, 51–55, 69–70, 72, 76, 79, 80
Disymmetry, molecular, 78, 79, 80
Double refraction, 55–58, 59, 60–61, 62, 64
Dumas, Jean Baptiste, 38–39, 40, 42, 44, 45, 46, 47, 48, 69, 73, 79, 80

École des Mines, 45
École Normale Supérieure, 7, 51, 68, 69, 70, 71, 75, 79
École Polytechnique, 35
Egyptian expedition of Napoleon, 19
Electrical polarity (boracite), 34, n.9, 52
Electrochemical dualism, 28, 38, n.31, 40–48
Electrolytic decomposition, 43
Elements, 10 (primary and secondary), 14, n.37, 50 (chemical)
"Elements" (Laurent), 50
Élie de Beaumont, Jean Baptiste, 40
Emerald, 28–29
Enantiomorphism, 5, 7, 33, 51, 69, 70, 73, 74, 75, 77–80
Equivalent weights, 50
"Essential" constituents of minerals, 18–19, 23, 29–30

Faculté de Medicine, 45
Faculté des Sciences, 51
Fermentation, 67
Fluorspar, 13, 34
Force (on light), 57–59, 63
Form-giving principle, 7, 10, 15
Foster, William, 35
Fourcroy, Antoine François de, 36
Freiberg, Mining School at, 14
Fresnel, Augustin Jean, 63–64, 70, n.11

Galena (lead sulfide), 43, 47
Gas hypothesis (Avogadro-Ampère), 35–41, 46, 50, 80
Gaudin, Marc Antoine, 37–41, 42, 46, 48, 54, 64
Gay-Lussac, Louis Joseph, 36, 45, 60, n.31
Geometrical chemistry, 26, 33, 35–37, 38, 40, 52, 79, 80
Gerhardt, Charles Frédéric, 50, 76
Germ (seed), 7, 10
Germ theory of disease, 80
Giobertite, 50
Glasses, 26, 60
Goniometer, 10, 29, 33
Gray copper, 23
Grignon, Pierre Clément, 10

Hankel, Wilhelm Gottlieb, 76
Haüy, René Just, 5, 11–14, 16, n.11, 17, 18–19, 20, 21, 23, 24, 25, 26–30, 31, 32, 33, 34, 35, 36, 39, 42, 45, 51–52, 54, 55, 56–57, 58, 60, 70, 74, 75, 78, 79, 80

81

Hemihedry, 52–55, 68–70, 71, 72, 76–79
Hemisomorphism (hemimorphism), 49, 51, 75
Herschel, John Frederick William, 62–64, 65, n.63, 70
Higgins, William, 22, n.4
Hooke, Robert, 7, 34, 35
Huyghens, Christiaan, 7, 34–35, 55–57
Huyghens's construction (for double refraction), 56–58

"Individual," mineral, 15–20
Iron pyrite, 12, 13, fig. 1, 53–54, 70
Isomerimorphism, 49
Isomerism, 33, 37, 41, 42, 48–49, 66–68, 71, 74–75, 76, 78
Isomorphism, 25–26, 29–33, 37, 41–43, 45, 48–51, 60, 68, 71–77, 79–80

Jardin du Roi (Muséum d'Histoire Naturelle), 16, 17, 20, 51

Kekulé, Friedrich August, 80
Kirwan, Richard, 14
Klaproth, Martin Heinrich, 23, 28, 30

LaPlace, Pierre Simon, Marquis de, 30, 57–58, 59
Laurent, Auguste, 7, 31, n.56, 37, 40, 41, 42, 43, 45–51, 53, 54, 55, 65, 68, 69, 71–75, 77, 78, n.55, 79–80
Lavoisier, Antoine, 6, 22, 45
LeBel, Joseph Achilles, 41, 80
Leblanc, Nicolas, 29
Liebig, Justus von, 45
Light, corpuscular and wave theories of, 55–58, 62, 63
Light—axis and poles (for double refraction), 57
Linnaeus, Carl, 9–10, 15–16, 80

Macquer, Pierre Joseph, 8–10, 21
Malus, Étienne Louis, 56–58
Mechanical intermixture, 29–30, 60
Micas, 60
Mineral taxonomy, 13–26, 28–32, 60; doctrine of fixed mineral species, 14–15, 17–26, 28–30, 32, 60
Mitscherlich, Eilhard, 30–33, 37, 42, 43, 48, n.89, 66–70, 76–77, 79
Molecular aggregation in crystal, 8, 10, 11, 59, 63–64, 68, 70, 74

Molecule, 7, 21, 27, 35, 39; 61–68, 73–74 (and optical activity); 34–54, 77–78, 80 models of atomic arrangement in); 37, 40, 42–43, 52–55, 59, 70, 79 (molecular form and its relation to crystal form)
Molécule, 36–37
Molécule constituante (elementary molecule), 11, 14, 26, 30–32
Molécule intégrante (integrant molecule, integrant particle, integrant part), 7–8, 10–14, 17–21, 23–28, 30, 32–33, 35–36, 42, 44, 52–53, 55, 79
Molécule organique, 15–16
Murray, John, 7, 14, 21

Naphthalene, 45, 48–49, 78, n.56
Naturphilosophie, 57
Newton, Isaac, 5, 55–57
Newtonianism, 57, 58, 59
Nicotine, 72

Paracelsian, 7
Paratartrates (racemates), 66–71, 73–78
Particule, 36–37
Pasteur, Louis, 5–7, 33, 51, 52, 53, 54, 59, 61, 68–80
Persoz, Jean François, 66–67
Physical optics, 6, 55–56, 61
Physical reality (of *molécule intégrante*), 12–13
Plesiomorphism, 54, n.118
Polarimetry, 72
Polarization, 6, 56–60 (and double refraction); 63 (circular); 33, 52, 58, 61–75, 78 (rotatory polarization—optical activity)
Polymorphism (dimorphism), 25, 30–32, 37, 41, 42, 59, 60, 66, 71, 74–75, 77, 79
Principle of least action, 58
Proust, Joseph Louis, 20–26, 28
Provostaye, Frederic Hervé de la, 48, 68–69, 73, 75
Pseudomorph, 29
Pyroelectricity, 70, 76

Quartz, 53, 58, 61–65, 69, 70–71, 74

Radical (nucleus, type), 42, 45–51, 54, 72, 75–80
Refractive index, 59
Robinet, Jean-Baptiste René, 10, n.16
Romé de l'Isle, Jean-Baptiste 7, 9–12, 14, 16, n.12, 17–18, 20, 29

Rouelle, Guillaume François, 8, 10, n.15, 15

Sage, Balthazar Georges, 9
Sal-ammoniac, 37
Saline principle, 9
Salt (sodium chloride), 43, 47
Scala naturae, 17
Seignette salt (double sodium-ammonium tartrate), 76–77
Silicates, 54, 78, n.55
Simplicity, assumption of, 12, 37
Sohncke, Leonhard, 35, n.11, 70, n.11
Solutions, 24, 60, 65–67, 73
Starch, 66
Stereochemistry, 5, 41, 80
Striation, surface, 52, 54
Strychnine, 72–73
Substitution, 38, n.31, 42–46, 48, 68
Sugar, 66, 67
Symmetry (molecular and crystalline), 40, 52, 53–55, 64–65, 69–70, 77

Tartrates, 5, 33, 65–71, 73–79
Thénard, Louis Jacques, 59
Thomson, Thomas, 20–21, 27, 33
Thomson, William (Baron Kelvin of Largs), 35
Tourmaline, 53
Transdiction, 6
Transparency, 13, 26, 60

Urea, 66

Vacua (between molecules), 13, 34
Valence, 41, 50, 54, 55, 80
Van't Hoff, Jacobus Henricus, 40, 41, n.44, 80
Vauquelin, Louis Nicolas, 28

Wallerius, Johan Gottshalk, 10, n.17
Water of crystallization, 37, 75–78
Werner, Abraham Gottlob, 14, 20, n.37, 23, 30
Wernerian mineralogy, 26
Williamson, Alexander Williams, 45
Wöhler, Friedrich, 45, 66
Wollaston, William Hyde, 13, n.35, 27, 29, 33–35, 37
Wurtz, Charles Adolphe, 41, n.44

Young, Thomas, 57